T0136079

About Island Press

Island Press is the only nonprofit organization in the United States whose principal purpose is the publication of books on environmental issues and natural resource management. We provide solutions-oriented information to professionals, public officials, business and community leaders, and concerned citizens who are shaping responses to environmental problems.

Since 1984, Island Press has been the leading provider of timely and practical books that take a multidisciplinary approach to critical environmental concerns. Our growing list of titles reflects our commitment to bringing the best of an expanding body of literature to the environmental community throughout North America and the world.

Support for Island Press is provided by the Agua Fund, The Geraldine R. Dodge Foundation, Doris Duke Charitable Foundation, The Ford Foundation, The William and Flora Hewlett Foundation, The Joyce Foundation, Kendeda Sustainability Fund of the Tides Foundation, The Forrest & Frances Lattner Foundation, The Henry Luce Foundation, The John D. and Catherine T. MacArthur Foundation, The Marisla Foundation, The Andrew W. Mellon Foundation, Gordon and Betty Moore Foundation, The Curtis and Edith Munson Foundation, Oak Foundation, The Overbrook Foundation, The David and Lucile Packard Foundation, Wallace Global Fund, The Winslow Foundation, and other generous donors.

The opinions expressed in this book are those of the author(s) and do not necessarily reflect the views of these foundations.

About SCOPE

The Scientific Committee on Problems of the Environment (SCOPE) was established by the International Council for Science (ICSU) in 1969. It brings together natural and social scientists to identify emerging or potential environmental issues and to address jointly the nature and solution of environmental problems on a global basis. Operating at an interface between the science and decision-making sectors, SCOPE's interdisciplinary and critical focus on available knowledge provides analytical and practical tools to promote further research and more sustainable management of the Earth's resources. SCOPE's members, thirty-eight national science academies and research councils and twenty-two international scientific unions, committees, and societies, guide and develop its scientific program.

SCOPE 68

Communicating Global Change Science to Society

SCOPE Series

SCOPE 1–59 in the series were published by John Wiley & Sons, Ltd., U.K. Island Press is the publisher for SCOPE 60 as well as subsequent titles in the series.

SCOPE 60: *Resilience and the Behavior of Large-Scale Systems,* edited by Lance H. Gunderson and Lowell Pritchard Jr.

SCOPE 61: *Interactions of the Major Biogeochemical Cycles: Global Change and Human Impacts,* edited by Jerry M. Melillo, Christopher B. Field, and Bedřich Moldan

SCOPE 62: *The Global Carbon Cycle: Integrating Humans, Climate, and the Natural World,* edited by Christopher B. Field and Michael R. Raupach

SCOPE 63: *Invasive Alien Species: A New Synthesis,* edited by Harold A. Mooney, Richard N. Mack, Jeffrey A. McNeely, Laurie E. Neville, Peter Johan Schei, and Jeffrey K. Waage

SCOPE 64: *Sustaining Biodiversity and Ecosystem Services in Soils and Sediments,* edited by Diana H. Wall

SCOPE 65: *Agriculture and the Nitrogen Cycle: Assessing the Impacts of Fertilizer Use on Food Production and the Environment,* edited by Arvin R. Mosier, J. Keith Syers, and John R. Freney

SCOPE 66: *The Silicon Cycle: Human Perturbations and Impacts on Aquatic Systems,* edited by Venugopalan Ittekkot, Daniela Unger, Christoph Humborg, and Nguyen Tac An

SCOPE 67: *Sustainability Indicators: A Scientific Assessment,* edited by Tomáš Hák Bedřich Moldan, and Arthur Lyon Dahl

SCOPE 68: *Communicating Global Change Science to Society: An Assessment and Case Studies,* edited by Holm Tiessen, Gerhard Breulmann, Michael Brklacich, and Rômulo S. C. Menezes

SCOPE 68

Communicating Global Change Science to Society

An Assessment and Case Studies

Edited by
**Holm Tiessen, Mike Brklacich,
Gerhard Breulmann, and Rômulo S. C. Menezes**

A project of SCOPE, the Scientific Committee on
Problems of the Environment, of the
International Council for Science

Washington • Covelo • London

Library of Congress Cataloging-in-Publication Data

Communicating global change science to society : an assessment and case studies / edited by Holm Tiessen ... [et al.].
    p. cm. — (SCOPE series ; 68)
"A project of SCOPE, the Scientific Committee on Problems of the Environment, of the International Council for Science."
Includes bibliographical references and index.
ISBN-13: 978-1-59726-176-0 (cloth : alk. paper)
ISBN-10: 1-59726-176-9 (cloth : alk. paper)
ISBN-13: 978-1-59726-177-7 (pbk. : alk. paper)
ISBN-10: 1-59726-177-7 (pbk. : alk. paper)
  1. Research—Latin America—Case studies. 2. Communication in science—Latin America—Case studies. 3. Research—Government policy—Latin America—Case studies. 4. Inter-American Institute for Global Change Research. I. Tiessen, H. (Holm)
  Q180.L37C65   2007
  303.48'3—dc22   2007014819

Contents

Part I: Crosscutting Issues

Part II: Background Chapters

Part III: Examples of the Science-Policy Interface

List of Figures, Tables, and Boxes

Figures

Tables

Boxes

Foreword

By Margaret S. Leinen and Paul E. Filmer

National Science Foundation
Arlington, Virginia, USA

Over the last two decades, governments around the world have begun to acknowledge that changes in the environment affect human activities in ways that are increasingly important. Supplies of essentials—shelter, water, and food—can be affected by decadal and longer cycles in climate and other Earth system processes, including disasters that are more intense, are more frequent, affect more people, or often all of the above. Governments are acknowledging that the environment is no longer a reliable invariant, resources are not inexhaustible, and we can no longer count on "business as usual."

Much financial support has been given to the scientific research community by governments to observe and investigate Earth processes, trends, abrupt events, and disasters. Governments are especially interested in what might be "tipping points" or bifurcations in chaotic systems like the climate. Fundamental scientific inquiry—discovery—will always remain one driver of such studies; however, *relevancy* has become increasingly important. Governments need Science (where Science is the larger, combined enterprise of the natural *and* social sciences) to actively contribute toward solutions to the problems caused by the complex, nonlinear interactions between the changing environment and our social structures.

Other communities are focusing on this need for integration of scientific research with the policy process. The international development community is concerned with sustainability of agriculture, water management, and desertification issues, as well as urban planning—all topics in which global change is a factor and to which Science should have significant input. In the nongovernmental world, conservation efforts are guided by greater awareness that ignoring climate change may make resource allocation decisions about particular locations and species moot.

National and international donors are also focusing on issues of global change. The coming decade of global environmental change research will result in increasing pressure to ensure connections with the decision needs of governments. Member states of the United Nations are striving to achieve the Millennium Development Goals, which are inextricably linked to closer integration of scientific information with policy development.

For many scientists and research institutions, this is a new context. It is also a new context for many science *funding* agencies, which have not previously had to manage the interactions between different areas of science and policy that this approach implies. There are very real needs for scientific input into the policy and research management worlds. An example of past successes is the 1987 Montreal Ozone Protocol. However, despite the significant difficulties already associated with the science underlying the negotiation of that protocol, the scientific community is well aware that the larger issue of global environmental change is much more complex than the ozone-related subset. Most of the simple Science problems have already been addressed.

The global environmental change science landscape is evolving rapidly. Earth systems science has done well in its transition to team-based work *within* Earth systems science, but it is just starting to tackle team-based work with social scientists and with nonscientists like policy and media specialists. Other disciplines, like engineering and medicine, have worked in this interactive mode for much longer (centuries in some cases) and will certainly respond to this new challenge effectively. The global environmental change science community and its institutions must adapt to this new mode of collaboration if they are to contribute to a sustainable future.

Future climate-related crises may actually *decrease* the amount of funding available for global environmental change science. Three factors contribute to this possibility: First, extreme events such as the hurricane-related damages on the Gulf Coast in the United States cause enormous reorganization in government funding structures, and the disciplines able to respond on short (and therefore perceived as *relevant*) timescales will generally benefit: medical services, social services, reconstruction logistics, and so on. The ability of Science to provide answers on timescales of hours, days, or even weeks is limited—the *immediate* value of the information currently provided is extremely low despite long-term payoffs. Second, extreme events are usually geographically restricted, compared with many global change issues. Paradoxically, regional and local crises can have greater impact on funding priorities than global ones, even if the local events are exacerbated by *global* change. A third factor contributing to the possibility that funding will decrease for global environmental change research is that this science is framed within fundamentally different timescales from those governing political and public interest. Without a concerted effort by the scientific community at consciousness raising and education on global environmental change issues, phenomena that occur on decadal timescales (not to mention century, millennial, or longer timescales) will have little traction in the political and popular world that determines funding.

If Science does not rise to the occasion, it risks becoming *irrelevant* in the policy decision chain addressing sustainability, and, possibly worse, the scientific community may lose ground in funding for all science efforts. Ensuring Science's relevance will require increased internal coordination between the natural and social sciences and their governance structures, as well as ongoing dialogue with important external social actors, including the policy sector, business, industry, and labor. These stakeholder interactions will require *adaptive management*, which has fundamental implications for all levels, including government funding agencies. Project management (by scientists and the grantee institutions, as well as by the overseeing granting agencies) will have to allow for flexibility in project objectives and for changes in the basic metrics used for scientific research and productivity. Budgetary allowances will have to be made to accommodate the increased need for science communications, additional dissemination modes, and interactions specialists.

The historic reluctance of science to engage in this process is, of course, related to the perceived trade-off between Science's long-term credibility and the relatively short half-life of policy interests. A carefully tended dialogue must be maintained with governments concerning what is currently "policy relevant" and what the related policy impact indicators are.

This volume explores the interactive frontier between Science and policy and looks at bridges between the two in the context of transformative research carried out in the Americas, primarily under the sponsorship of the Inter-American Institute for Global Change Research (IAI), an international treaty organization. In spite of the difficulties in organizing new research networks, there are several successes in the included chapters that demonstrate that the interactions described above between science and policy can indeed change local and national government policies. Interestingly, one of the IAI's greatest achievements has been the formation of a cadre of academic administrators in the region who are now capable of administering multi-institutional, multicurrency, and multidisciplinary research projects and who are now pursuing funding from regional development agencies.

The processes of policy making and of science almost always differ from our preconceived notions—rather than being linear processes that lay out a question, analyze it, and propose a solution based on that analysis, both science and policy making require complex, nonlinear, iterative processes that deal with multiple, interlinked, and changing questions. Closer cooperation between two such processes cannot be expected to be simple. All actors must learn something about the peculiarities of the others' culture.

We look forward to that conversation.

Acknowledgments

This book, the workshop that generated the crosscutting chapters, and the Collaborative Research Networks (CRNs) that were the source of all but one of the case studies in the book were funded by the CRN I grant (ATM 9907233) to the Inter-American Institute for Global Change Research (IAI) from the US National Science Foundation.

Véronique Plocq-Fichelet, the executive director of SCOPE; Susan Greenwood-Etienne from the SCOPE Secretariat in Paris; and Luis Marcelo Achite, Roseli Luz, and Isabel Vega de Maldonado from the IAI provided essential support and guidance before and during the workshop. The staff from the Recanto das Toninhas Hotel in Ubatuba created the excellent working environment in which the crosscutting chapters were written. Florence Carmont prepared the manuscripts for submission to the publisher.

The science-policy analysis conducted in this book has resulted in the UNESCO-SCOPE Policy Brief No. 3, published in December 2006.

John W. B. Stewart, Editor in Chief

SCOPE Secretariat
51 Boulevard de Montmorency, 75016 Paris, France

1

Why This Book?
An Introduction and Synthesis

Holm Tiessen

There has always been global change, and global change has always had impacts on human populations. Twelve thousand years ago, hunters migrated out of the drying Sahara region, and northern people colonized the fertile soils left behind by retreating glaciers of the last ice age. Only a hundred years ago, some of these fertile soils were settled by mostly European agriculturalists who had migrated to North America, many of them driven by droughts in the southern part of the Soviet Union. For much of human history, people have reacted to climate change by adapting to or escaping from environmental stresses and exploring opportunities. On today's more populous and wealthier planet, opportunities for escape have become limited, while increased knowledge and systematic scientific approaches have improved the chances for adaptation. At the same time, humans themselves have become a significant cause of global change. This is the root of an increasing demand for global change science to 1) predict the rate, shape, and extent of global change; 2) provide decision aids for mitigation; and 3) provide guidance toward adaptation.

In response to these demands, science is undergoing transformations toward greater societal and policy relevance, both in its choice of subject matter and in its communication of results. These transformations do not happen in a linear or planned process but occur in an undirected manner as scientists and research institutions respond to changes in science funding, attitudes, and policies. This book presents both a collection of experiences from the Collaborative Research Networks (CRNs) of the Inter-American Institute for Global Change Research (IAI) and an analysis of the policy interface that has developed in these networks. Researchers prepared background chapters and provided interviews about the policy relevance, links to decision makers, and outreach of their projects. Some projects had no policy component. These typically claimed an intrinsic policy relevance based on the project's scientific excellence and importance, but no activities were undertaken that might have conveyed this importance to society or policy makers. Most research networks started out without policy agendas but developed them during the course of the five- to six-year projects as opportunities arose. The development of policy relevance was often the result of exposure of the researchers to societal needs during the research. Only one project was fully embedded in government policy; it was based in government administrations and dealt with public health issues due to global change. Despite this integration, the implementation of research findings in public health practice was difficult because practitioners found it difficult to integrate the predictions and probabilities of an epidemic based on climate events into their day-to-day operations.

Both development and economic growth are undermined by global change. Therefore any response to global environmental change requires political, as well as scientific or technical, treatment. Under the realities of global change, societies must ask, Development and growth of what, for whom, and at what cost? "Green" policies are often challenging to the policy sector because they typically call attention to what should *not* be done and therefore often emphasize the "negative" side of "positive" development and growth. Legitimacy of such policies is founded on the need to harmonize development with the capacity of Earth systems to support societies' needs; therefore legitimacy is founded on scientific knowledge. The policy dilemma results because democratic decision makers usually arrive at policies by an adversarial process, while on global change issues, they are compelled to exercise persuasion and inducement in a continuous learning process linked to science. In this delicate situation, in order for science to succeed in influencing policy and the decisions or actions of societies, it has to have a number of attributes that go beyond traditional measures of scientific quality. These needed attributes are *credibility, acceptability, practicality, usefulness,* and *accessibility.*

Scientific *credibility* is underpinned by peer reviews of proposals, research methods, and results. Wider credibility is commonly derived from endorsements by sources or authorities trusted by society or its representatives.

Credibility and *acceptability* are enhanced when policy makers are engaged in the research process, starting with the initial framing of the research questions. Their continued involvement requires that scientists be ready to engage in communication and remain responsive to demands by the policy sector. It is a difficult lesson for many scientists that, to be policy relevant, good science is not enough but a persistent and patient engagement with the appropriate audience for the scientific message is also required. The message may also have to be translated and packaged for different audiences. Such processes have been the hallmark of commissioned work. Yet global environmental change (GEC) science cannot be driven only by articulated needs of societies. Research must also be proactive, ahead of societies' demands.

The development of policy relevance has enhanced many scientists' engagement and integrated them into ongoing communication with different stakeholders. This process requires safeguards against biases, as the development of advocacy may undermine the neutrality demanded of the scientific process. One finding of the research on climate risk is that climate stress does not cause human vulnerabilities but only unveils them. Close cooperation between natural and human sciences is required to establish full causal chains between natural events and human conditions. This interdisciplinarity has been one of the most difficult and least successful tasks of the research networks. Where it did occur, interdisciplinarity often meant adding a discipline to the research process, not really integrating it.

If scientists themselves want to influence policy, they have to understand the relationship between policy and politics. One of the most important components of that understanding is an appreciation of who will win and who will lose as a consequence of the results of their science.

To influence policy, scientists have to identify the right person to hear the message at the right time, and in the appropriate language. Often this person is hard to find, the timing is hard to establish, and the language is hard to learn. To ease links to society and policy relevance, several of the projects have found "brokers," most often in the form of nongovernmental organizations (NGOs). NGOs involved in the search for development alternatives or in resource management have a need for scientific information. In turn they have offered scientists access to communities or research sites. They have also carried the knowledge or technical solutions provided by the scientific research back into the communities, using communication and demonstration tools with which scientists are less conversant. The cooperation has often been fruitful and has developed its own momentum, which has frequently continued after project funding ceased. Yet, in the beginning, there were very significant barriers caused by perceptions of different agendas and lack of transparency and understanding of goals and motivations of different actors. Essentially, trust needed to be built before the groups could work together effectively. This took time and effort; it also sometimes required adjustments in research agendas.

A recurring issue is that the tasks of such engagement and communication are not recognized in traditional scientific or academic reward systems. Even work involving cooperation between different science disciplines is often seen as a dilution of scientific rigor. The typical merit system based on peer-reviewed publications in disciplinary journals does not value interdisciplinary work to the same degree, so such work counts less toward promotion and recognition.

Practicality and *usefulness* have not been highly valued criteria unless they have resulted in patent rights. Throughout the consultations that have led to this book, we have heard calls for making adjustments in the academic reward system and the criteria for funding and monitoring success.

Accessibility in the scientific world means open data sources, accessible metadata, and publications in recognized journals. Accessibility for society and decision makers implies condensation, digestion, and translation of scientific results—all tasks for which scientists are ill trained and rarely rewarded. Typically only late-career scientists take on these tasks, because they have overcome the career-determining pressures of the academic reward system.

In addition, scientists may be discouraged by the variable and shifting interests and motivations of stakeholders and by their intermittent engagement. It is important for the GEC science community to recognize that other stakeholders may engage in only those phases of the science process in which they have sufficient interest. In turn, discontinuous engagement by scientists is also an obstacle to stakeholder engagement. Land users typically have time horizons well beyond the duration of research projects. When stakeholder engagement during research generates interest and further questions, scientists may no longer be able to respond when project funding comes to an end. Despite the benefits of longer-term engagement, the search for innovative portfolios may make funding agencies reluctant to make longer-term commitments.

Despite the many obstacles, most researchers in the IAI projects evaluated for this publication not only established policy links and stakeholder involvement but also found the process highly rewarding, particularly in the international and intercultural settings provided by Collaborative Research Networks. Often students and young scientists were involved. There was an expectation in the projects that young researchers trained in this environment would engage in policy linkages as a matter of course in the future. This was seen as a very important aspect of IAI networks. There is the hope that this investment will last for years through a generational change in which trainees will be both scientists and policy makers of the future.

One motivation for the present analysis is that it will be useful in the development of future GEC research programs. Lessons from the analysis translate into an evolution of the tasks of research-funding agencies. New funding programs have to create opportunities (funds that allow connecting what has not been connected so far) and challenges (progress beyond business as usual). Meeting standards that combine the criteria for both policy relevance and scientific excellence places greater professional and personal demands on the researcher. Such projects not only focus on the generation of scientific knowledge (as in the traditional science-funding approach) but also require additional investments in time and human and institutional capacity building. This brings both opportunities and transaction costs. The development of this type of project is a dynamic process with mutual feedbacks, readjustments, and uncertain outcomes. Programs must be steered with more caution and greater effort from both the researchers and the funding agencies. The project cycles likely affect both researcher and funding agency. This new process may be as important as its outcome, since it presents a great opportunity for institutional development and learning.

There is much talk of internationalization of academic institutions and their science, often accompanied by little action. The Collaborative Research Networks had to work internationally as a condition of funding. Despite the higher transaction costs of international projects, the intercultural and international work proved useful for the science outcome. In one project, Brazilian researchers participated in interviews of land users in Peru and Ecuador, and vice versa, as a means to offset the common tendency of researchers to assume that what they know about their own country equally applies to others. The mixing of interviewers from different nationalities also contributed to the comparative analysis by sensitizing researchers to similarities and differences that would otherwise have been overlooked.

The word *stakeholders* appeared in many analyses of policy relevance. Earth Summit 2002 defined stakeholders as "those who have an interest in a particular decision, either as individuals or representatives of a group . . . [including] people who influence a decision, or *can* influence it, as well as those affected by it." This broad definition includes everyone related to a research project: its participants, its objects, and even its funders. Often the roles were changed by or during a project. In one case, project stakeholder engagement was secured through subcontracting only. In another, stakeholder engagement was triggered by a convincing environmental threat but rapidly waned when economic concerns became more prevalent in people's minds. The need for a perceived threat makes early stakeholder engagement in GEC science difficult since the science may be needed first to describe the threat fully. The divide between investigator perceptions and public or national policy agendas inhibits early engagement. On the other hand, NGOs may share agendas and may stay engaged with both policy and science.

Many of the case studies presented in this book are anecdotal. Some stand out for the poverty of their engagement and insight. In their entirety, though, the studies characterize the imperfect process of developing policy relevance, links to society, and stakeholder involvement in GEC science. The GEC science community is an exponent of the process of developing policy relevance because its science is increasingly driven by societal concerns. The critical analysis of the case studies is based on background chapters on communication, institutions, societal vulnerability, interdisciplinarity between social and natural sciences, and international conventions as legal frameworks. This analysis is presented in these crosscutting chapters: Steering Research toward Policy Relevance; Stakeholders and Global Environmental Change Science; Delivering Global Environmental Change Science to the Policy Process; and Communicating Science to the Media, Decision Makers, and the Public. The many and varied experiences and their analyses provide valuable lessons for improving the process of including societal and policy relevance into GEC science and science in general. The process is certainly needed as societies demand returns for their investments into a science that is needed to solve critical problems of societies and their well-being in a changing world.

PART 1
Crosscutting Issues

2

Steering Research toward Policy Relevance

Mark D. Stauffer, Walter E. Baethgen, Ricardo L. Berbara, Ernesto Caetano, Alejandro Castellanos, Barbara Göbel, Michael E. McClain, Rômulo S. C. Menezes, and Arturo Sanchez-Azofeifa

This chapter addresses the challenge of making global change research relevant to policy makers, that is, influencing local and/or national policies as they are being configured. We approach the interface between science and policy from a specific angle: the experience of the researchers involved in Inter-American Institute for Global Change Research (IAI) projects. We are not taking into account the perspective of the other stakeholders, and therefore this chapter is not intended as a systematic assessment of the policy impact of the research projects. However, valuable insight was garnered by this process of review, analysis, and synthesis. Interviews with all principal investigators revealed varying degrees of commitment and means to achieve policy relevance in the research projects even though relevance was not thoroughly planned (i.e., appropriate policy development expertise was not included). The lessons learned should help both the IAI and its Collaborative Research Network (CRN) principal investigators achieve greater policy relevance.

What are the "policy-relevant" responsibilities of the IAI? The IAI is based on the mutual agreement among 19 nations that a mechanism is needed to generate knowledge regarding environmental changes and to inform nations about the dynamics of global change phenomena so that the affected nations may develop policies responsive to expected change. The IAI has responsibilities for the following: 1) identifying relevant contemporary research themes, 2) financing and managing scientifically credible projects to generate knowledge about these themes, and 3) reporting back to the science community and member countries the results of the research. The IAI also has responsibilities for capacity building and supporting the internationalization of science in the

Americas. These actions are policy relevant at a second order in that they increase the capabilities of nations to conduct the scientific research necessary to guide their decisions.

Policy-relevant research yields results that are of specific use to decision makers regarding global change issues. The research should also be supportive of education and capacity building. The decision maker may be the president of a country or the head of a single household. IAI projects should strive to provide results that are as useful as possible to national actors (in public and private sectors) and that support decisions at as high a priority as possible. It is a fundamental mandate of the IAI that its science should be policy relevant.

The IAI fulfills its responsibilities aided by its structure. The first responsibility is met through formulation of science themes. The second responsibility is met through the implementation of science programs. The third responsibility is met through its science publications and information exchange across the member countries. Investigators supported by the IAI also have policy responsibilities. They should ensure that 1) their research addresses high-priority issues and generates results that are of use to decision makers, 2) their results are of high scientific quality, and 3) their results are made available to the appropriate decision makers in an appropriate format.

The IAI has a multiple-step evaluation process that serves as the first guidance for steering research toward policy relevance. Evaluation criteria are clearly described in each call for proposals. The first and most important criterion is scientific excellence. Proposals are peer reviewed in a three-stage process by researchers with expertise in the field(s) of the proposal. After scientific excellence is considered, attention is given to IAI goals such as interdisciplinarity, links between natural and human sciences, and regional distribution of participating countries and institutions. The IAI can, at this point, help ensure positive outcomes regarding policy relevance of the research supported.

Dimensions of Policy Relevance in IAI Projects

Neither the IAI call for proposals nor the design and initiation of research projects address policy relevance in an explicit or narrow way, restricting it, for example, to certain types of stakeholders and institutions operating at a specific level (local, regional, national, or transnational) or to certain types of science-policy interaction (e.g., science-based assessments or policy briefings). This relative freedom relies on the idea that the research will find its natural outlet and that its relevance will flow to the appropriate recipients so that policy—either formal or informal—will result. But is it sufficient? Researchers of the IAI projects have interacted with community authorities, representatives of regional and national governments, nongovernment organizations (NGOs), the business sector, and international organizations in the environmental and developmental arenas. Some of these stakeholders and institutions have operated in quite isolated ways; others have been parts of broader networks linking several scales. Many of the science-policy interactions have been meetings, face-to-face communication, information exchange and discussion, or written reports with resulting recommendations. Only few research projects have had the financial and human capacities to deliver a broad range of materials, such as school books or regular policy briefs. Most projects have moved beyond one-time interactions with stakeholders and built up long-term relations based on mutual interests, trust, and predictable outcomes. This indicates that first-order science-policy interactions have occurred. In many cases, it could be argued that greater value to the research would be achieved if a higher-order interaction followed so that the policy developed at one level would be complemented by recognition and/or action at other critical levels of policy and so, in fact, consolidated. The questions then become these: To what extent can the IAI drive the science-policy agenda? What is the cost, and is it affordable? And, equally important, does policy development expertise exist?

Common Patterns of Project Development and Implementation with Regard to Policy Relevance

There is no linear relation between science and policy. Instead, different pathways with different costs and benefits for the actors involved exist. These pathways have rarely been analyzed in a systematic way in order to provide a better basis for decisions by researchers with regard to policy relevance. CRN researchers interviewed by the authors of this chapter described the policy relevance components in their projects as anywhere from "no policy component" to "no policy agenda, but it developed during the course of the project" to "an opportunistic approach" for connecting the research to decision-making processes and embedding it in government policy. Hence a common strategy was to plan for the unplanned. By being sensitive to the possibility that a policy issue could arise, policy-relevant opportunities could be captured within the ongoing research agenda. Policy relevance was also present in projects that supported broad government goals, specifically those related to capacity building through training or meetings in the countries as well as formal training abroad. A more detailed analysis of the IAI projects revealed two basic pathways for policy relevance in the development and implementation of research agendas.

Pathway One

Many IAI projects started with a research agenda exclusively based on the knowledge of the participating researchers of the region and their expertise on the targeted research problems. Only later, in the course of project development, was policy relevance built in and made explicit. Several factors could trigger the inclusion of this additional dimension to the scientific agenda—funding requirements or the direct exposure of the researchers to societal needs during the process of data collection, for example.

Each group of researchers followed one of two processes to include the policy dimension. Some groups organized workshops attended only by researchers in order to specifically discuss stakeholder issues and thereby include those issues into the proposal. Other groups gave the stakeholders a more active role by inviting them to participate in workshops that captured issues for inclusion in the proposal. The IAI Start-Up Grant Program, which facilitated network building, proposal development, and initial research, was central to catalyzing both processes. The researcher-stakeholder interactions served several interrelated purposes:

- readjustment of the research agendas in the sense of a "reality check," that is, confronting the researchers not only with other views of the problems but also with different prioritizations of the problems
- collection of additional information to make more-complete pictures of the problems and their contexts
- embedding of themes and problems in broader economic, social, political, and cultural contexts

Pathway Two

Some IAI projects started because stakeholders, mainly national or regional institutions or international organizations, defined concrete demands for scientific information or for science-based solutions. Research groups picked up these demands and translated them into research agendas. Often further interactions with the stakeholders were used to readjust the agendas and steer their implementation. The early involvement of stakeholders created awareness on the part of researchers to generate policy relevance through the proposed research.

It was stressed that researcher-stakeholder interactions in both pathways created additional knowledge, reshaped "traditional" scientific knowledge, set the basis for the translation of knowledge, and triggered a process of mutual learning. It should also be noted that the role of researchers, as well as that of stakeholders, could change in the course of interaction. For example, the group of researchers could become the mediator between local farmers and regional or national governments to convey to the respective parties their sets of needs and concerns. In other cases, stakeholders mediated between researchers from different regions and disciplines. From the perspective of the researchers, most concerned with the needs and constraints of their research projects and the characteristics of the knowledge production process, it was extremely important to choose the right stakeholder at the right moment for the right role in the process. It was found that, normally, this happened on an ad hoc basis and, as a result, created high transaction costs.

Many researchers found that by involving a mediator such as an NGO with a long history of work in a region, they were able to present their research "product" in such a fashion that it met the needs and demands of end users such as local farmers. It was also indicated that other relevant actors such as experts, consultants, or knowledge brokers in the science-policy interface should get involved as multipliers. As scientists are not policy makers, such multipliers are in a better position to organize, from an epistemological and an organizational point of view, the bridging between science and practice.

This synthetic analysis of past IAI projects provided some general pointers for the process of building policy relevance into a research project:

- The uncertain dynamics of the process and its "fuzziness" must be expected.
- Continuous readjustments are needed and important.
- The ill fit between the expectations of the different actors involved may be presumed to be inevitable at some level but should be discussed in the early stages of the project preparation or implementation.
- A diverse set of different mechanisms of engagement (e.g., workshops, publications) will likely be revealed as the project unfolds and will need dynamic orchestration.

The outcomes of the interactions between researchers and stakeholders must be reinforced. For example, outputs could be added such as the following: databases or other archives of knowledge; communication, outreach, and dissemination strategies; and education on different levels.

Successes and Failings in the Pursuit of Policy Relevance in the Collaborative Research Networks

The CRNs investigated a broad range of global change phenomena with relevance to national and local decision making in the Americas. The policy relevance of the CRNs took different forms, ranging from direct implications for decisions on specific problems to more indirect outcomes such as increasing the pool of qualified global change scientists and enhancing awareness, communication, and international networking among actors in the global change community. All projects aimed to quantify and describe phenomena of global change, and thus to clarify the situation a country or region faces, as well as to provide information in a context from which decision makers can act. Some projects also addressed more-specific issues and problems, yielding results that could be translated directly into policy recommendations.

The scientific output of the CRN program was a considerable success, as demonstrated by the many scientific publications in peer-reviewed books and journals. Success in capacity building and networking is also clear: The CRNs provided graduate training for hundreds of scientists, and hundreds of community members were involved in the CRNs' larger networking and outreach activities. Success in terms of *pol-*

icy relevance is more difficult to assess, because it cannot be easily quantified. If sound science findings are policy relevant, the unstated question largely centers on how to get information to flow to those who need it. A subjective measure of success in the policy relevance of CRNs may be the degree to which project knowledge and/or recommendations are incorporated into policy or are assimilated into political or practical thinking. A more indirect measure of success may be the number and nature of contacts established between the researchers and the policy community, extending from the planning phases to conclusion of the project. Since there was no systematic recording of measures of relevance within the CRNs, the only source of information is the recollections of the investigators.

The fundamental policy relevance of the CRNs was established by the four IAI science themes in which they were framed. These themes were identified through a collaborative process involving scientists and policy makers of the region. Individual projects addressed elements and subsets of these themes. For example, several projects addressed issues related to land use change, especially its impacts on soil fertility, ecosystem function, and water balances. Many of these projects included parallel studies to understand land use changes in the context of economic and demographic changes and climate variability. Other projects focused on understanding the dynamics of climate change processes and natural variability. These generated data that can be used by other investigators to examine processes and infer policy implications. Among the most directly relevant projects were those that focused specifically on the risks to human health and welfare that are linked to global change.

Across the CRN program, a great deal of emphasis was placed on understanding the consequences of change and especially the impacts on ecosystem services. Thus CRN research was often comparative and often capitalized on the concept of substituting space for time. The chronosequence concept enabled investigators to assess potential future changes to ecosystems by comparing coexisting systems. For example, comparison of processes in forests, pastures, and agricultural fields was used to assess the impact of land conversion on ecosystem processes.

In several countries of the Americas, the success of the CRN program can be measured simply by the initiation of applied global change. Government budgets to support global change research are nearly nonexistent in a number of IAI countries. Even in countries such as Mexico and Argentina, the contributions of the IAI significantly enhanced national programs in global change research by providing longer-term funding. Thus the CRN program established, for the first time in several countries, a community of scientists to serve as counterparts to and collaborators with those policy makers addressing global change issues. In some instances, CRN fora became venues for discussing global change issues. In Mexico, natural and social scientists active in the CRN program responded directly to requests from the Mexican administration for information on land use planning and afforestation. The information they provided prompted Mexican authorities to allocate resources to respond to some of the problems posed by climate change. Authorities in Mexico and other countries of the region that support mitigating initiatives are beginning to understand the need to generate knowledge regarding the advantages and disadvantages of certain provisions or statements generated by the international Climate Fora.

The advanced modeling applied in several CRN projects also added a level of precision and sophistication to the dialogue on global change in several countries. Examples can be seen in Peru, Ecuador, and Mexico. In Mexico, after some years of collaboration, the interaction between the Ministry of the Environment and CRN scientists has resulted in an initiative to create the Mexican Program on Climate Modeling and Uses of Information. This project has resulted in a science plan, which is currently evolving into an implementation plan. Another example is the joint project of the National Autonomous University of Mexico, the University of São Paulo, and the University of Costa Rica on numerical models for seasonal predictions of regional climate. It has resulted in new capacity in Mexico and Costa Rica to prepare climate information. Continued collaborations with governments of the region should lead to the use of such information in regional planning.

Obstacles to developing science-policy interface stemmed mainly from a lack of clear understanding of expectations and of the strategies and levels of effort needed to see the process through. The pursuit of policy relevance is a new concept to many scientists, and there is no commonly shared understanding of the meaning of *policy relevance*. The CRN program is novel in its requirement of multinational collaboration. As investigators noted, the implementation and maintenance of effective local and international networks greatly increased their workload during the first years or even throughout the project. Having collaborators from different backgrounds and institutions with different agendas required great effort to establish communication (meetings, workshops, traveling, etc.). The time and energy invested into these activities detracted from more-common scientific and academic pursuits and traditional performance indicators. Investigators also pointed out that research activities often took place far from the political centers of the countries and often addressed themes that had not yet arrived on the national agendas. This physical and thematic distance from national policy makers ultimately resulted in stronger bridges being built to local policy makers than to national partners, as originally envisioned. In two CRN projects, the regional issues addressed were not (yet) a priority for policy makers at the national level, given the distance (in all meanings) between the region of study and the federal government policy makers. When IAI science does not match with national or regional policies, two different messages may be conveyed: Either scientific knowledge is ahead of the policy process and should play a role in educating the policy community, or the science agenda is not properly situated and not effectively communicated to the target audience.

The pursuit of policy-relevant research within the CRN program was loosely structured, in the sense that policy relevance was required of projects but no systematic guidance was provided on what it constituted or what mechanisms should be designed into projects to optimize policy outcomes. CRN investigators interpreted for themselves the meaning of policy relevance and built bridges of their own design to the policy community. Consequently a number of strategies were developed. The successes and weaknesses of these efforts provide valuable lessons that can be applied in improving future projects. With time, greater guidance could be provided by funding agencies such as the IAI. Expectations should be clearly articulated, and recommendations should be given for project design.

Lessons Learned in Steering IAI Scientific Programs for Policy Relevance

To qualify for IAI funding, research proposals must comply with IAI's objectives:

- scientific excellence
- multinational involvement
- multidisciplinary collaboration
- contribution to capacity building
- policy relevance and usefulness to stakeholder decision makers and policy makers

These defined prerequisites increase the chances that IAI-funded research will achieve policy relevance. Both establishing multidisciplinary teams and stimulating stakeholder involvement in research programs whose funding periods are longer than the one or two years typical in Latin America help a CRN program to achieve policy relevance. The required internationalization of the proposals increases the capacity of scientists to participate in international science. Moreover, the internationalization of the research provides opportunities for framing program agendas in a much broader international context. In some cases, results of CRN projects are direct inputs for international initiatives and conventions.

This combination of requirements makes the IAI research program different from traditional scientific funding mechanisms. The innovative approach implies opportunities (funds that allow connecting what has not been connected so far) and challenges (progress beyond business as usual). The funding criteria place greater professional and personal demands on the researchers, since the focus of these projects is not only on scientific knowledge production, as in the traditional science-funding approach, but also on additional investments in time and human and institutional capacity development.

The development of this type of project is a dynamic process with mutual feedbacks, readjustments, and uncertain outcomes. Programs of this type must be steered with more caution and greater effort from both the researchers and the funding agencies. In some respects, this new process is as important as the outcome since it presents a great opportunity for institutional and epistemological learning. One challenge in this novel process is to ensure that scientific excellence is not compromised.

Costs/Risks and Benefits of Including Policy Relevance

The experience of principal investigators and co-principal investigators of all CRNs was that including policy relevance in their research brought about great benefits but also significant transaction costs and risks. A commonly mentioned benefit of making the research policy relevant was the opportunity for participants to attract new funding (from national or international sources) based on the work they were conducting under the CRN. The required active interaction of scientists with stakeholders in some cases allowed them to witness and even steer the impact of their research in the policy process.

On the other hand, the investigators mentioned the high transaction costs of seeking policy relevance in their research. For instance, new and different pressures were placed on academic institutions and funding agencies to broaden the range of capacity building to include school teachers, farmers, practitioners, or government representatives. As a result, the traditional concept of capacity building in research programs, that is, increasing the scientific capacities of participating individuals and research institutions, had to be expanded. Finally, it should be mentioned that there is often concern on the part of scientists that they risk their credibility by being too concerned about policy relevance, particularly when they must move beyond their scientific expertise into areas where they have no training.

Importance of Temporal and Spatial Scales and How to Overcome the Mismatch between Scales

One of the outstanding features of the CRNs was that participating scientists and stakeholders collaborated at different spatial scales. Collaboration among scientists was at the regional level (different countries in the Americas). Scientists also interacted with the global environmental change research community, contributing to international and global initiatives. Collaboration between stakeholders and scientists typically occurred at the local level and sometimes occurred at the national level.

An issue that the CRN investigators identified as essential for attaining policy relevance was the inclusion of the right stakeholders as partners in their activities. The experiences of the CRNs indicate that successful stakeholder involvement goes hand in hand with a high degree of community involvement and with power linkages within the geographical and societal scale of the scientific project. The CRN also revealed that effective strategies for stakeholder involvement often include working with different stakeholders at different stages and phases of the project.

Oceanographic and atmospheric projects with focus at national and regional, rather than local, scales were able to attract the attention and support of national governments. The Brazilian Navy supported the research of one of these CRNs with ship time beyond the national waters (Campos and Piola, chapter 14, this volume). A different situation was found in another CRN, whose research was conducted in somewhat peripheral semi-arid environments not suited to high-input agriculture (Menezes et al., chapter 16, this volume). In this case, research results had direct impact mostly at local scales within the study areas. Yet lessons from these studies were also applied in other regions with similar characteristics in the involved countries. Other examples of regionally limited policy relevance were found where research results, although relevant at local scales, were difficult to scale up to the national or regional level. For example, research conducted in human health problems by one CRN produced knowledge that could be applied at very local scales (Cofalonieri, chapter 22, this volume), while policy makers usually seek information that assists their work at much broader scales.

Necessary Follow-Up Mechanism to Guarantee Sustainability of the Investments into Policy Relevance of Research

One main concern expressed by investigators was that, with the termination of the CRN project, all the efforts invested in networking, collaborating, and establishing formal working relationships with stakeholders often also ended. It is important that projects plan for continuing involvement (and funding) long before the project cycle ends. This is a shared responsibility of researchers and funding agencies. Continuous efforts during project execution are required to develop the trust and credibility of the research within the policy community, and that trust can be lost very rapidly in the absence of follow-up activities.

Final Remarks

Over the past 10 years, IAI projects have included policy relevance components that have proved to be useful and important to the member nations, because they induced scientific endeavor to bear upon social issues and provide answers to decision and policy makers. Although the process for making research policy relevant was "fuzzy," in many projects, opportunities were seized and the research was found to have good utility in both policy and practice. The CRN projects did not involve, at either the project level or within the IAI, the process or expertise required for adequate planning with regard to policy relevance. However, innovative thinking, flexible action, trial and error, and—importantly—researcher determination to seize opportunities produced effective information, activities, and networks that were ultimately policy relevant.

It is suggested that more formal guidelines or instruction on how to generate and steer research toward policy relevance be generated from this experience. Funding agencies should make it clear to those submitting proposals that policy relevance is an important criterion for project selection and that the proposal should outline means by which policy relevance is to be achieved. In addition, opportunities to make research policy relevant will arise, often unexpectedly, and they need to be captured; hence, researchers should plan for the unplanned.

International collaborations improve the capacities to promote the use of scientific information in the planning or development process across countries with different abilities to conduct or lead scientific initiatives.

The review of IAI-supported projects over the past 10-year period identified the need for an expanded role of funding agencies, one that strengthens the researchers' ability to develop and maintain linkages with governments and other major stakeholders. Funding agencies may be in the best position to explain the research they support and to place it into societal context, that is, to mesh research into the existing body of knowledge as well as into the concerns of nations.

3

Stakeholders and Global Environmental Change Science

Mike Brklacich, I. Foster Brown, Edmo J. D. Campos, Alex Krusche, Allan Lavell, Kam-biu Liu, Juan J. Jiménez-Osornio, Susanne Reyes-Knoche, and Charles Wood

Human impacts on the environment have been studied for more than a century, but it is only in the past 25 years that it has become widely recognized that human activities are transforming Earth on a global scale. Fundamental alterations to Earth system processes (e.g., climate systems) and the cumulative effects of many smaller changes in aggregate contribute to the sweeping degradation of environmental resources and services on a global scale (e.g., conversion of coastal marshes). The rise of global environmental change (GEC) science has also required fundamental shifts in science processes. Isolated experiments and local observations simply do not provide the necessary perspective, and it is now common for science to be mobilized around international research teams employing common protocols for the development and sharing of globally constructed databases. This maturation of GEC science has also posed new challenges for the social sciences, as it is now clear that altered environmental systems are impacting human well-being at local to global scales and that GEC solutions will require imaginative, comprehensive, and integrated efforts. Human activities are GEC drivers, and solutions to GEC concerns will need to be socially constructed.

The breadth of GEC science has resulted in an unprecedented need for collaboration between the natural and social sciences, and the range of participants in these processes has been expanding rapidly at many levels. "Think globally, act locally" efforts have expanded participation in GEC science to include representatives from governments, nongovernmental organizations (NGOs) and civil society organizations (CSOs), resource users, and educational communities. This expanded research base has attracted multiple funding agencies from the private and public sectors (Jutro, chapter

7, this volume). The GEC research family now includes many stakeholders who are either directly involved in the GEC science process or affected by GEC.

The term *stakeholder* is employed in its broadest sense in this chapter. The origins of the term can be traced back to its application in the financial sector, where a stakeholder is a third party who serves as a caretaker of funds that are of joint interest to two or more parties. In that context, the stakeholder is a neutral and trusted representative of all interested parties. Contemporary usage in many settings, including the applied sciences, has expanded the term's definition from that narrow financial application, and the term *stakeholder* is now commonly employed to denote all parties with a voluntary or involuntary legitimate interest in a project or entity. Earth Summit 2002 defined stakeholders as "those who have an interest in a particular decision, either as individuals or representatives of a group . . . [including] people who influence a decision, or *can* influence it, as well as those affected by it." This definition emphasizes the plurality and differing motivations for stakeholders who have now become integral parties to GEC science. Hence the full constellation of stakeholders includes not only GEC scientists, who are actively engaged in the execution of scientific projects, but also the following:

- other scientists who participate in GEC science but are not actively involved in the project
- GEC science funding agencies
- a wide range of interest groups such as national and regional policy agencies, NGOs and CSOs, and resource managers who are involved in the project either directly (e.g., data collection, preparation and vetting of reports) or in the steering of the science process
- individuals and communities that are vulnerable to GEC
- interest groups and individuals who either endorse or oppose the GEC science project

There are many reasons for multiple stakeholders' participation in GEC research. For example, early stakeholder integration facilitates the development of research questions and products that are meaningful to both the users and the generators of GEC science. In addition, multiple domains of knowledge, many beyond the realm of the science community, can provide insight into and solutions for GEC issues. This is only possible when stakeholder integration is founded upon mutual trust among all participants in the GEC science process.

The purpose of this chapter is to provide insight and guidance on stakeholder motivation and participation in GEC science, as well as conclusions aimed at guiding the participation of a wide range of stakeholders in GEC science over the next decade.

Lessons from the IAI CRN Experiences

It is not surprising that an international program as varied as the Inter-American Institute for Global Change Research (IAI) research initiative involved a large and diverse number of stakeholders. This chapter draws on interviews with seven of fourteen principal investigators (PIs) of the first round of IAI's Collaborative Research Network (CRN). These interviews were conducted in December 2005 as the CRN projects were approaching completion. The PIs provided information on the kinds of stakeholders involved in their networks, when the various stakeholders became involved, the roles they played in the process, and their motivations for participation. It is important to note that this chapter is based on the PIs' assessment of stakeholder involvement in GEC science projects. The direct consideration of the stakeholder communities that participated in these GEC projects is beyond the scope of this chapter.

Some of the studies focused on a strictly defined scientific issue. For example, one CRN (Nuñez, chapter 15, this volume) investigated climate variability and change in the Mercosur region and, as a result, initially involved only the team of GEC scientists; stakeholder involvement came late in the process, at the communication and dissemination stages. Another CRN (Wood and Tourrand, chapter 17, this volume) assessed the role of cattle ranchers in deforestation and land use change in Brazil, Ecuador, and Peru and drew on participants or targeted audiences and interest groups from the outset. It therefore engaged stakeholders at many points in the research process. As a result of the broad portfolio of CRNs, from studies of basic Earth system processes to assessments of GEC impacts on human activities and well-being, no single model for engaging stakeholders emerged. Instead, many different and legitimate strategies were employed to expand stakeholder communities beyond the IAI CRN teams and to engage these stakeholders in the GEC science process.

A conceptual framework for organizing and understanding the complexity of stakeholder involvement (Figure 3.1) can be organized around six interrelated science activities, from defining the research questions to the communication of research results within and beyond the science community. Interactions between these activities and the multiple stakeholders may be mapped on this framework, in which the scientists perform the IAI project and each group of stakeholders has its own "card in the deck" to represent its involvement in the science process. The intent of the figure's three-dimensional feature is to capture the notion that multiple stakeholders can be involved at various points along the six stages of the process.

The "conventional science" trajectory can be depicted in Figure 3.1 as A1, A2, A3, A4, A5, A6. That is, all stages in the GEC science process, starting from defining the research questions all the way through to the communication of scientific findings, conventionally fall within the domain of a single group—the project scientists—which is really only one of many constituencies comprising GEC science stakeholders. "Participatory science" lies at the other end of the involvement continuum and provides shared responsibilities between the project scientists and all other stakeholders (other scientists, policy analysts, resource managers, etc.) at each stage of the process. Its trajectory in Figure 3.1 is A1/B1, A2/B2, A3/B3, A4/B4, A5/B5, A6/B6.

The majority of IAI CRN projects employed intermediate strategies, in which par-

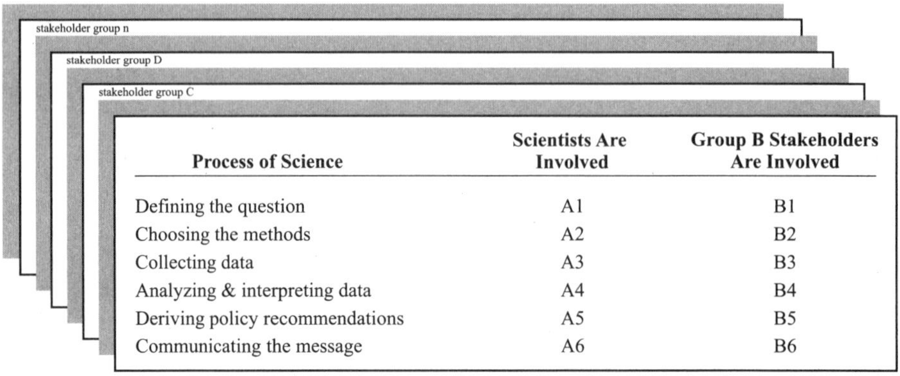

Process of Science	Scientists Are Involved	Group B Stakeholders Are Involved
Defining the question	A1	B1
Choosing the methods	A2	B2
Collecting data	A3	B3
Analyzing & interpreting data	A4	B4
Deriving policy recommendations	A5	B5
Communicating the message	A6	B6

Conventional Science Trajectory: Typically only scientists are involved
A1 => A2 => A3 => A4 => A5 => A6

Participatory Science Trajectory: Scientists and stakeholders are involved at all stages
A1/B1 => A2/B2 => A3/B3 => A4/B4 => A5/B5 => A6/B6

Common Intermediate Trajectories
A1/B1 => A2 => A3 => A4 => A5 => A6
A1 => A2 => A3/B3 => A4 => A5 => A6
A1 => A2 => B3 => A4 => A5 => A6/B6
etc.

Figure 3.1. Conceptual framework for organizing and understanding the complexity of stakeholder involvement

ticular stakeholders entered and exited at various stages of the research process. For example, planning agencies helped define a river management study in the Amazon and helped communicate the project's findings (McClain et al., chapter 19, this volume). Relationships between these two groups (i.e., the IAI-sponsored science team and the planning agencies) occurred during the first and sixth step in the process, so the trajectory was A1/B1, A2, A3, A4, A5, A6/B6.

Key Findings and Lessons Learned

A large variety of stakeholders participated in the CRNs. Who they were, how they became involved, and how they stayed involved was examined in detail.

Stakeholders Participating in GEC Science: Who They Are and Their Motivations

The CRN experiences showed that there is no single reason or motivation driving stakeholder participation. This is consistent with studies elsewhere, which also conclude that stakeholder participation in the environmental sciences is highly variable (Stoecker 1999; Mitchell 2002). The role stakeholders play depends on the nature and interests of the group in question and on the stage in the process at which it becomes involved. A study of the chemistry and biology of the southwestern Atlantic Ocean (Campos and Piola, chapter 14, this volume) relied on naval ships from two countries to collect data (A3/B3 in Figure 3.1). However, it was apparent this support would not be available until preliminary findings from the project were made available beyond the immediate GEC science community. It was the discovery of the common interests between the two stakeholders that facilitated the direct involvement of the navies from two countries in that CRN. A study of river hydrology in the Amazon (McClain et al., chapter 19, this volume) established a relationship at the data collection phase (i.e., A3/B3). In this case, the stakeholders were school children who conducted routine water-level measurements. Diverse as the stakeholders may have been—national navies on the one hand, and children in local communities on the other—their participation in both cases assisted with data collection efforts and this, in turn, provided the CRNs with opportunities to raise the profile of GEC issues in these stakeholder communities and to develop scientific capacity at the individual and institutional level.

Factors motivating the IAI-sponsored scientists to expand their research networks beyond their immediate team were also varied and related directly to the overall purpose of their research projects. In some cases, systematic assessments of a specific GEC topic were unavailable or there was insufficient infrastructure in place to support GEC research (Luckman, chapter 12, and Confalonieri, chapter 22, this volume). The concept and structure of the CRN provided a means to establish and coordinate a larger and more integrated scientific consortium. For other projects, motivating factors included the desire to engage with nonscientific stakeholders who had a vested interest in GEC science but traditionally had relatively undeveloped links to the GEC science community (Wood and Tourrand, chapter 17, Klink et al., chapter 18, van den Belt et al., chapter 20, this volume). For those CRNs where there was a significant resource management component (Menezes et al., chapter 16, and Lavell, chapter 21, this volume), broadening of the stakeholder communities involved in the project provided a mechanism to share project findings with groups that were not directly involved in the project.

Engaging Stakeholders in GEC Science

The CRN experiences also illustrated there was no single or optimal time when stakeholder involvement should commence and terminate. In most cases, stakeholder involvement was constrained to a few of the six steps in the scientific process illustrated in Figure 3.1. The initial defining of the scientific questions commonly involved only the CRN-sponsored scientists. Other stakeholders entered the process in subsequent phases, most frequently during the data collection and analysis phases. In some cases, stakeholders then continued throughout the investigative process, but more commonly their involvement was only temporary. Given that interests and motivations among the various stakeholders often are quite varied, it is more likely that intermittent engagement will prevail. Hence it is important for the GEC science community to recognize that stakeholders can be active players at any and in multiple phases of the scientific process. Key challenges include identifying those phases in which there is sufficient common interest to warrant collaboration and to not insist upon continuous engagement as a condition for participation in a GEC science project.

Managing and Maintaining Stakeholder Participation in GEC Science

The IAI CRN experiences demonstrate that collaborations among GEC scientists and stakeholders, including NGOs, government agencies, educational institutions, and donors, are dynamic processes that can enhance the science process and outcomes (McClain et al., chapter 19, this volume). This engagement of multiple stakeholders, however, creates new expectations among the various stakeholders, and it is not without costs. The CRN experiences have revealed several lessons for managing multiple stakeholder processes and for maintaining stakeholder involvement.

It is key to recognize the diverse opportunities and situations that prompt and benefit from multiple stakeholder engagement. For some stakeholders, motivations are tied directly to their professional responsibilities (Campos and Piola, chapter 14, this volume), whereas the motivations of other stakeholders are embedded in broader educational goals (Luckman, chapter 12, and McClain et al., chapter 19, this volume) and in community-based environmental management (Lavell, chapter 21, this volume). This engagement must be flexible to accommodate contributions at various stages throughout the scientific process. The search for common themes among the stakeholders provides a foundation for collaboration, and the management of multiple stakeholders can benefit from a careful articulation and communication of this common ground. While shared interests between stakeholders may provide a basis for collaboration, this does not imply that stakeholders must agree in world views, values, or backgrounds. Therefore the management of multiple stakeholders demands communication strategies that are tailored to specific stakeholder needs in order to avoid conflict among the stakeholders. All stakeholders must be encouraged to respect other points of view, especially in terms of understanding and resolving the needs and problems faced.

Stakeholders as Agents for Capacity Building and Change

The stakeholders of the IAI CRN initiatives range from the typical—regional scientific communities, graduate students, and government agencies—to a diverse assemblage in which farmers, fishers, ranchers, local politicians, military forces, teachers, local NGOs, health professionals, and tourist guides participate in or are affected by the science process. Many of these stakeholders, such as university professors and teachers, have capacity building as a primary function (Luckman, chapter 12, this volume), but capacity building occurs in many other venues as well. In several CRN initiatives, these education-oriented stakeholders served as direct multipliers for the transfer of research results to a wider audience. For example, elementary school teachers explained to school children the monitoring of local rivers in the Andes of Peru, and university professors gave a short course in Brasilia, Brazil, to those interested in carbon sequestration projects. Others, such as extension agents, provided training for farmers in agricultural methods that help sustain soil fertility and economic productivity. *Capacity building* refers to educational programs at all levels as well as extension programs.

The capacity building in itself served to expand the perspectives of stakeholders and enabled them to understand better the context of GEC science of the CRN initiatives. At the same time, it provided a means for two-way communication that enabled researchers to interpret the interests of other stakeholders and their needs for information. In the case of studies in the semiarid regions, where land users have to adapt to adverse climate, this feedback helped to shape the design of the research.

Capacity building in many of the CRN initiatives increased the impact of the science and also provided a venue for dialogue between stakeholders and the scientific team that improved the research and its applicability for policy at local and national levels.

Evaluation

Stakeholder participation was not a major consideration in most CRN projects' design and selection. Although somewhat vague references to policy relevance were included, neither definitions of stakeholders nor relevance were offered. Therefore an evaluation of these factors and their importance, successes, and problems is both subjective and based on hindsight. Stakeholder participation in the science process is obviously relevant, but given the problem of varying definitions of stakeholders, from the science community itself to the policy-making community and affected population groups, any analysis or conclusion must be considered tentative and exploratory. Yet one may derive the recommendation that future projects make such processes explicit and that IAI design a qualifying methodology and define impact indicators. Nevertheless, analysis of the CRN science process points to the importance of several elements, including the following:

- networking in research and the incorporation and stimulation of interest on the part of diverse stakeholders
- greater participation in "hard" science research projects by social science actors who can help develop policy and questions that are relevant to society, and "humanize" research
- stakeholder participation in order to improve data and information collection and analysis and subsequent interpretation and application
- local-level research and analysis and the participation of local stakeholders, who are more likely than national and international actors to sustain participation

Emerging Issues

The GEC science landscape is evolving rapidly, as is the context for stakeholder participation in GEC science. This section summarizes key issues concerning stakeholder participation in GEC science that are beginning to emerge and will need to be addressed by the GEC science, funding, and policy communities over the next decade.

International Environmental Conventions

In many GEC projects, stakeholder involvement is based on the motivation of the scientists in charge. The scientists make the choice of stakeholder involvement according to their needs (scientific questions) and evaluation of the different phases and aims of the project. Depending on the research field, international conventions and the corresponding national legislations can require the involvement of different stakeholders (Reyes-Knoche, chapter 10, this volume). For example, the Convention on Biological Diversity (CBD) recognizes the sovereign right of states over their biological resources. This sovereign right of states—reflected in national laws—encompasses the right to grant or deny access to their genetic resources for scientific activities. Hence GEC scientists have to observe the international and national regulations regarding access procedures that define the kind of stakeholder involvement required. Closely linked to this are the regulations concerning the sharing of benefits resulting from the utilization of genetic materials, as well as regulations governing stakeholder participation, transfer of technologies, the sharing of results from scientific activities, and capacity building. International conventions or national laws have important stipulations that can set the context for stakeholder involvement in GEC research. It is crucial that these factors be considered as part of the project's design when key scientific questions are being defined, and also later during the project's execution.

Social Learning and Stakeholder Involvement

GEC science requires institutionalized interdisciplinary teams that are academically solid and include multisectorial participation. The teams must be capable of organizing training methods that facilitate the sharing of results, create accessible alternatives for comprehensive training while ensuring its continuity, and identify organizations and institutions that are committed to learning together.

Stakeholder involvement offers an outstanding opportunity for interactive learning through exchange of knowledge and experiences via formal or informal education. It is a two-way learning process between the different network components, and it will have impact not only on individuals but also on institutions. This facilitates capacity building among different stakeholders and can have significant impact on policy.

Adaptive Research Design—Changing/New Stakeholders

In a commentary published recently in *Nature*, Patrinos and Bamzai (2005) argue that the path between climate science and policy is not always linear. This argument could be extended easily to stakeholders and GEC in general. This implies that a GEC scientific project should follow an adaptive research design, which would respond to the input and/or feedback from stakeholders throughout its execution. The IAI CRN experience shows that stakeholders are involved in different phases of and with a variety of effects on the scientific process. There are situations when they are integrated at the onset of the project, taking part in its initial definition. In other cases, when stakeholder involvement occurs in a later phase, the integration of the new stakeholders may require some reorientation of the science plan in order to accommodate their needs and contributions.

Integrating New Stakeholders

In the context of the IAI, the integration of new stakeholders must be set within the realities of the Latin America and the Caribbean (LAC) region. Of the half billion people in LAC, 30 percent are under 15 years of age, and a quarter exist on less than two dollars a day. By 2050, it is estimated, the region's population will grow 44 percent to 800 million. Clearly, one major group of GEC science stakeholders in the region is the young and poor. This reality underscores the urgent need for curriculum reforms that ensure that GEC issues are routinely incorporated in the classroom. Recent activities of the Large-Scale Biosphere-Atmosphere Experiment in Amazonia (LBA) in Acre, Brazil, illustrated one approach for curriculum reform when LBA researchers provided rural schools with background information on global environmental change as well as satellite images of recent deforestation. The opportunities for this kind of engagement are immense where over 100 million people in LAC are of school age. In Brazil alone, 55 million children are in the basic education system and represent a significant portion of the population that is most vulnerable to GEC.

Large infrastructure projects, such as the Initiative for the Integration of South American Regional Infrastructure (IIRSA, http://www.caf.com) and the interoceanic highway from central Brazil to Peruvian seaports, also generate a new demand for stakeholder involvement. These and similar projects have become key GEC drivers as they accelerate deforestation, becoming corridors for migration and economic restructuring. As a result, they expand the range of stakeholders who are exposed to GEC. In many cases, these infrastructure projects propel local communities into a GEC context and thereby bring local professionals, NGOs, government officials, and community leaders into the GEC community. In addition, these large projects often transcend national political boundaries and can become catalysts for the formation of new stakeholder coalitions such as the trinational MAP Initiative, the alliance of three local administrative units in Peru (Madre de Dios), Brazil (Acre), and Bolivia (Pando).

Stakeholders and Traditional Knowledge

The involvement of local or indigenous communities in several aspects of GEC research is increasing, especially in the areas of environment, biodiversity conservation, and its sustainable use. *Traditional knowledge* (TK) remains a contested term. For an overview of the different approaches to define *traditional knowledge* and *indigenous knowledge*, see the World Intellectual Property Organizations (WIPO) 1998–1999 report. Article 8j in the CBD defines TK as the knowledge, innovation, and practices of indigenous and local communities embodying traditional lifestyles relevant for the conservation and sustainable use of biological diversity (CBD 2005). The definition is flexible and thereby provides involved parties (the states) with the mandate to protect and safeguard traditional knowledge within their respective national realities.

Informed consent—based on provision of the project's aims and structure, as well as understanding of the benefits and risks arising from participation in the project and how they will be shared—safeguards the participation of TK holders. In addition, TK usually is owned not by a single person but by the whole community, and this can complicate the decision about which parties should be involved in a GEC project. While one community member can be assigned as the key link between the community and the GEC project, it is important that benefits are shared with the whole community. Key factors to guide the TK applications in GEC projects include the following:

- Traditional organization and decision-making structures of these communities must be respected.
- Relevant local communities should be consulted during the conception of the project and in its diverse stages, and informed consent should be integrated into the project from the outset.
- In the case of research activities related to access to genetic resources and associated traditional knowledge, the dialogue with the communities should include transparent negotiations about intellectual property rights.
- Should the project generate new costs to support capacity building or additional assessments by the local community, then the project's principal investigators need to assist with securing additional funds to maintain local contributions.

The Media as a Stakeholder

The mass media (daily and weekly press, popular science magazines, radio, television, and some cinema) play an important role in molding, directing, forming, and deforming public opinion and, collaterally, influencing policy change and formulation (Weingart, chapter 6, this volume). The development of current important themes, topics, and concerns and the presentation of scenarios, outcomes, and change options are common in the media. This goes beyond the sensationalist approaches of much of the popular press and covers important themes in a serious and accessible way. Increasingly GEC topics and concerns have made headway in the media, including discussions on global climate change and warming, increases in hurricane incidence and strength, desertification, sea-level rise and glacier and polar ice melting, deforestation, and land degradation. These are dealt with in ways commensurate with the media audiences and following different ideological and political guidelines or orientations.

GEC research and the research process, from knowledge generation to policy influence, compete with everyday, "normal" environment and development concerns for media attention. Experience suggests that both the public and politicians exposed to such coverage will assign greater priority to current issues and discount the importance of longer-term concerns, which are often couched in the context of uncertain future scenarios, so the GEC science community must learn to frame its products and messages in a more media-friendly manner. Global change researchers and institutions must be willing to learn how to enlist media support in fostering an understanding of GEC problems and in justifying needed policy options and change. The media can and should be a powerful ally of GEC research, but this will require efforts to facilitate access to the various media and to ensure GEC messages are transmitted effectively to targeted interested populations, policy agencies, and pressure groups.

Unequal Capacities

The CRN experiences showed that GEC science will routinely involve stakeholders drawn from multiple political and socioeconomic contexts and having different capacities to participate in the process. Hence the resources required for stakeholders to effectively participate in and comprehend GEC science and its human implications will also vary form location to location. The GEC science community must take into account that immediate needs for global action in many cases will differ from local interest and capacity to participate. Research in this field should address such heterogeneity through training and education and, whenever possible, should fund and support initiatives that could lead to a more equitable consciousness of local, regional, and global implications of GEC. Economic, social, and cultural differences should be considered as part of the scientific approach, in order to motivate stakeholders to the benefit of GEC science.

Maintaining Stakeholder Participation and Commitment

Relationships between stakeholders and GEC scientists are often intermittent, and therefore it should be anticipated that stakeholder interests will surge and ebb throughout a long-term project. The GEC science community will need to devise strategies to maintain stakeholder interest, especially during periods of less active participation. Such periods can lead to lower levels of participation in the long term and the eventual disengagement from GEC science. As a minimum, follow-up communications such as project briefings and press releases can be used to maintain commitment. Even though new stakeholders can be integrated, it is important to maintain stakeholder interest and commitment, not only during the project's execution, but also beyond its completion. Hence it is crucial that the plans for maintaining stakeholder involvement after the initial funding period be incorporated into the project's initial design.

Contextualizing Stakeholder Involvement

Contextualizing refers to the norms that govern GEC research in each country and the impacts, both positive and negative, on GEC science and stakeholder participation. Given that IAI's charter requires that funded projects be multinational and multidisciplinary, it is crucial that its projects be sensitive to the broader issues occurring within each participating country. Information gleaned from questionnaires completed by the PIs as part of the program evaluation provided evidence of the degree to which the character and content of the researcher-stakeholder relationships were shaped by the contingencies presented by particular research sites. In some countries, the media were receptive to the GEC message and became allies in disseminating information to audiences and interest groups, but this was not the case across the board. Similarly, interest groups affected positively or negatively by GEC research findings bring to the policy discussion different degrees of power, legitimacy, and effectiveness, which vary from one national context to the next. There is a need to properly "contextualize" the way in which researchers involve or engage audiences, participants, and interests groups in the science process.

Concluding Comments

Stakeholders are heterogeneous groups representing multiple interests in GEC science. There is no single reason motivating stakeholder participation in GEC science, and the stakeholders' roles are contingent upon their specific interests and capacity to participate. Resource managers with subnational mandates and educators at many levels are keen to participate in GEC science and become users of GEC science outputs. A key challenge for the GEC science community is to explore individual stakeholder interests and ensure that the common ground needed for participation is well prepared from the outset. When needed, the GEC science community has the responsibility for securing additional resources required for stakeholder participation.

Stakeholders choose to participate in various stages of the scientific process and seldom participate in all stages of a GEC project. GEC scientists need to be sensitive to stakeholder interest in the project and also be aware when stakeholders require science products. For example, the development of national and international environmental policies routinely involves multiple phases over lengthy periods. It is crucial that GEC scientists provide products in a timely manner that is synchronized with the policy process.

Stakeholders' participation in GEC science needs to be founded upon a mutual understanding of their contributions to the project and the benefits they will derive. Stakeholder participation in a GEC science project hinges upon mutual trust. Should this trust be brought into question, then future participation will be at risk. Hence it is crucial that expectations, especially of the role stakeholders will play in the project and the benefits they will derive from the project, be discussed openly and agreed upon by all parties at the project's outset.

Stakeholders make multiple contributions to GEC science, ranging from establishing the research agenda to participating in data collection to capacity building. Stakeholder participation needs to be cast in a broad context. Involving stakeholders at the earliest stages has the advantage of establishing a research project that will suit stakeholder needs and timing. Stakeholder involvement in data collection facilitates the completion of the project, but it can also enhance project impact through a sense of ownership. Stakeholders can facilitate dialogue with many other consumers of GEC science and thereby assist with identifying information needs of a broader range of stakeholders, as well as improve the applicability in related policy and resource management communities.

Stakeholder participation in GEC science must be in accordance with international and national law as well as consistent with local norms for the sharing of knowledge and benefits. GEC science often contributes to international environmental conservation and protection initiatives, and stakeholder participation in GEC projects must conform with international and national legal requirements, including the full recognition of the rights of sovereign states over environmental resources. In addition, the benefits derived from participation in a GEC project are often owned by a community rather than only the individuals who actively participate in the project. It is crucial that sovereignty concerns and community ownership be included in the project's overall design and during the execution of the project.

Stakeholder participation needs to be a planned set of activities within the GEC science process and be based upon an adaptive research design. Stakeholder participation in GEC science must not be an ad hoc activity. It is a purposeful set of activities designed to facilitate mutual benefits among all stakeholders. A project design based on early stakeholder participation that is also responsive to evolving interests throughout the course of the project will be required to maintain meaningful stakeholder participation.

GEC science must avoid overtaxing stakeholders and recognize that stakeholders have unequal capacities. Stakeholder interests outside the GEC project will compete for their attention and time. Therefore parameters governing stakeholder participation need to be carefully explored and established early in the project. Various stakeholders bring different skills to the project. The project design must ensure realistic goals for stakeholder involvement.

The GEC science community has a responsibility to maintain and manage an environment that fosters long-term stakeholder participation. Managing diverse interests is key to maintaining stakeholder involvement throughout and beyond the project. A project communication plan that keeps all stakeholders current on project activities will encourage continued contact with the project, even during project phases that are not of immediate interest. In addition, it is incumbent upon the GEC community to avoid conflict among stakeholders and assist with addressing the needs of all stakeholders.

Literature Cited

CBD. 2005. *Handbook of the Convention on Biological Diversity* (including its Cartagena Protocol on Biosafety), 3rd ed. Montreal: Convention on Biological Diversity. (http://www.biodiv.org/convention/convention.shtml)

Mitchell, B. 2002. *Resource and environmental management*, 2nd ed. Essex: Longman.

Patrinos, A., and A. Bamzai. 2005. Policy needs robust climate science. *Nature* 438(17): 285.

Stoecker, R. 1999. Are academics irrelevant? Roles for scholars in participatory research. *American Behavioral Scientist* 42(5): 840–54.

WIPO. 1998–1999. *Intellectual property needs and expectations of traditional knowledge holders: WIPO report on fact-finding missions on intellectual property and traditional knowledge.* Geneva: World Intellectual Property Organizations. (Publication Number 768) (http://www.wipo.int/tk/en/tk/ffm/report/index.html)

4

Delivering Global Environmental Change Science to the Policy Process

John Ingram, John Stone, Ulisses Confalonieri, Theresa Garvin, Peter R. Jutro, Carlos A. Klink, Brian H. Luckman, Elke Noellemeyer, and Peter M. de Toledo

There is an increasing demand for global environmental change science to contribute toward solutions as the stresses imposed by human activities on the life-support systems of planet Earth become more evident. One important mandate of the Inter-American Institute for Global Change Research (IAI) is to make a contribution to the development and application of science to the policy-making process in the region and in its 19 member countries. The IAI's scientific agenda was designed to be dynamic, allowing for the periodic incorporation of new scientific priorities while, at the same time, being policy relevant by addressing the needs of the region's countries. The purpose of this chapter is to distill some lessons from the experience of the IAI Collaborative Research Networks (CRNs) and from other international projects in the region, so as to build upon and improve the dialogue between scientists involved in the IAI programs and policy makers in member countries. This chapter will provide practical guidance to the science-policy process.

In this chapter, we have defined *policy* broadly to include all decisions that affect or are affected by global environmental change. The chapter will focus mostly on policy makers in governments, although the lessons learned can be applied to other actors, such as individual resource managers. The chapter recognizes that science is only one input to the policy-making process and that there is a distinct difference between policy and politics. It also recognizes that decisions often have to be made with incomplete or imperfect information.

The science-policy interface is not always transparent, operates at several levels, and involves many different actors. There are two basic elements to the policy process: governance and dialogue. The first determines the roles of the various actors. These include scientists; nongovernmental organizations; the media; lobbyists or others with vested interests; and decision makers and policy makers at local-, regional-, and national-level governments and the private sector. In principle it is to the advantage of the functioning of the science-policy process if the roles of each are well defined and understood. The field of global environmental change cannot be divorced from values. The experience of being involved in developing and applying scientific information can lead to being involved in the decision-making process. Nevertheless, it is imperative that scientists communicate their results and insights in as objective a manner as possible.

Ideally, effective dialogue within the policy process requires an open and transparent environment and the development of a common framework and language between each of the actors. In many cases, dialogue does not evolve spontaneously as an efficient, product-oriented process. The dialogue can, however, be improved when key attributes relating to the science and its presentation are recognized. This chapter derives and uses a set of necessary attributes based on the experience of the scientists involved in several CRNs. It is based on a systematic analytical approach that was designed to elucidate good examples of "what worked well" in relation to a given attribute. These are presented as a series of examples to facilitate an improved dialogue in the future. The returns from the considerable investment required in achieving an effective dialogue can have benefits that go beyond the initial motivation.

Necessary Attributes of Science for Policy

Not all science sets out to be policy relevant. Much fundamental research is not expected to be of immediate value to the policy process. If, however, science outputs are intended to influence the policy process, they need to have a number of key attributes. The experience of the CRNs has been used to establish these necessary attributes. These are *credibility, practicality, utility, accessibility,* and *acceptability*. While different situations may necessitate a particular emphasis on some attributes, all are important.

Credibility

The peer review process is the long-standing, principal way of establishing credibility of science outputs within the scientific community. This aspect is often central to the development of policy processes (e.g., in the Intergovernmental Panel on Climate Change—IPCC), and indeed some policy can be based only on science that has been peer-reviewed (e.g., in developing environmental legislation for the United States by the Environmental Protection Agency). In dealings with policy makers on a more personal basis, the notion of trust is also important, whether it be between institutions or individuals. Researchers must build and maintain the trust of all sides involved in the issue. Trust takes time to build, but it is fragile and can be rapidly eroded. The reputation of the scientist or institution is also important in developing this trust. Reputation is enhanced by demonstrated evidence that the scientist appreciates what is considered relevant to the user. This, in turn, helps to generate confidence on the part of the users, especially when reinforced by a transparent process of research agenda setting and interpretation of results.

Establishing Credibility

The work of one CRN exemplifies some of the ways credibility may be achieved in bringing science to the policy process (Luckman, chapter 12, this volume). Most of the scientific effort of this CRN was dedicated to the collection of data that could be used to document and understand local, regional, and hemispheric climate variability. These data sets are archived in free and open international databases, thus allowing for use and verification by the scientific community. Similarly, the results are published in reputable, peer-reviewed, international journals. These data sets can provide critical inputs to policy development and decision making at scales ranging from local water management to providing input to the IPCC.

As an example, credibility was enhanced at a national level in this CRN through an interdisciplinary expert group consisting of university scientists and professionals from forest companies and led by a Chilean co-principal investigator (Co-PI). The group developed a document: *Key scientific components for a national policy on the uses, services and conservation of the Chilean native forests* (Lara et al. 2003). This document and its recommendations were later endorsed by national-level groups composed of university and policy experts and presented to the Agriculture Commission of the Chilean senate and to the president. The document includes key recommendations for policy and legislation in Chile and is expected to have a significant impact on future decisions regarding native forests in Chile.

In another example, at a more local level in Mexico, results from a study of biomass yield from semiarid mesquite scrublands had immediate applications for the sustainable management of the wood resource. These results were included in a pamphlet distributed by the government forestry agency to producers as a guide to the future restoration, harvesting, and management of these resources.

Practicality

The ability to use scientific information to formulate and implement policy based on it depends on a number of aspects related to its practicality, that is, how easy it is to use to inform policy. First is the timeliness of information delivery in relation to need. Delivery may come about opportunistically or strategically in relation to upcoming policy formulation. Second, the information needs to be delivered in appropriate formats so that the users can readily grasp the key aspects without a "translation" process, which takes both time and effort (thereby reducing the likelihood of uptake) and may allow misinterpretation of key results (leading to frustration and erosion of the science-policy relationship). Third, as the policy process is driven by many factors in addition to science, the practicality of information also depends on its ability to be used in the context of other inputs. These will likely vary in scales of both time and space, which the science output may have to match. Offering a range of options may help to deliver science at relevant scales, especially in association with supporting databases.

Benefits of an Opportunistic Approach

Following a field trip to the experimental sites located in the Cerrado savannas near Brasilia, Brazil, a window of opportunity was grasped by the scientists of one CRN when they spotted a chance to present their results to officials from the World Bank and the Brazilian Ministry of the Environment (Klink et al., chapter 18, this volume). CRN scientists were engaged by the World Bank to prepare a report on analysis of stakeholders for the Cerrado and to prepare a concept paper on Cerrado land use and conservation in order to seek support from the Global Environment Facility (GEF). The authors were able to incorporate the knowledge generated by the CRN on biodiversity, water balance, and the carbon cycle and the description of key stakeholders and their spheres of influence. The overall suggestion from stakeholders and institutions was that a strong case could be made for the Cerrado conservation, given that a high level of habitat and landscape modification and degradation had already occurred. The report also enabled the World Bank and government officials to develop policy trade-offs between ecosystem functioning and economically valuable agricultural production that culminated in the approval of the Sustainable Cerrado Initiative by the World Bank, GEF, and the Brazilian Ministry of the Environment, thereby adding value to the basic research and increasing its practicality.

Utility

The usefulness of scientific information depends on the recognition of its value by the user. Time and effort will probably be required on the part of the science community to develop effective communication skills and help the users understand the value of science to the policy issues at hand. This is related to format but also to the legitimacy of the information in relation to the institutional context. These factors come together to affect the ability to act on information, although this can be compromised by the lack of willingness to act. The utility of the wealth of science outputs from the Large-Scale Biosphere-Atmosphere Experiment in Amazonia (LBA) is being increased by 1) integrating science information to respond to policy-relevant questions set by governments and 2) creating an appropriate government-sponsored extension to LBA (GEOMA—the Environmental Modelling Network of the Amazon created in 2002 by the Brazilian Ministry of Science and Technology).

Increasing Utility of Science Information

The evolution of the LBA project and the GEOMA network is an example of how science can produce useful, high-quality information on the environment in the Amazon region that is of value to policy makers and enhances the public perception of the importance of science. This has been achieved by the development of products, such as scenarios, that integrate monitoring information, knowledge of ecological processes, and human dynamics, as well as modeling. LBA/GEOMA is a response to the national and international interest in the Amazon forests and has provided results on ecosystem functioning, greenhouse gas emissions, climatology, and land use changes in the rain forest. The original large multilateral LBA experiment produced many scientific papers and built local scientific capacity, but it was not effective, on its own, in developing a formal dialogue between scientists and decision makers at the government level. In mutual recognition of the value of such a relation, GEOMA was established to produce commissioned products such as reliable scenarios regarding social and environmental impacts in the Amazon. Technical reports, oral presentations directly delivered to ministers, and participation in governmental working groups dealing with economic development of the Brazilian Amazon region have been some of the strategies used by the GEOMA network to deliver the results and diagnoses at a faster pace to decision makers. Environmental and social issues such as forest conversion, climate changes, and biodiversity losses have been shown to be of great interest during the building of developmental agendas and strategies in northern Brazil.

Accessibility

The accessibility of science to the users is often a communications issue. Appropriate language is required, both in terms of local language per se and in terms of the degree of science jargon, as users often are not trained in the science concepts. The clarity of information is also important, particularly in describing and communicating the concepts of uncertainty and complexity. Appropriate formats are needed, which may include highly visual material. This will help in the comprehensibility of the information. Language appropriate to the cultural context is a necessary condition for increasing the accessibility of scientific information.

Increasing Accessibility through Appropriate Communication

The most important challenges of the CRN on land use change in semiarid regions were related to accessibility (Menezes et al., chapter 16, this volume). This involved generating scientific information that could be used by farmers, translating this into practicable and viable land management options, and effectively communicating it. Addressing these challenges required an open process in which both scientists and rural communities learned from each other and found common terms and understandings. One difficulty was often language itself—in some rural areas in Yucatán, Mexico, people speak only Mayan. This difficulty was overcome by involving scientists who belonged to the indigenous culture of the study region and who were familiar with the idiosyncrasies of the local population.

This CRN also faced the problems of finding adequate approaches to communicating alternative management options as well as the complexity of the systems studied. Conducting experiments on farmers' lands to generate scientific information proved ideal for communicating with farmers. Through this approach, scientists were able to share and discuss their observations and results, evaluate the various implications of different management options, and facilitate the understanding of the complexities involved. As a result, adoption of the proposed management options proved to be very effective. This CRN was successful in overcoming the problem of accessibility, as farmers are now able to use the knowledge that was developed and are better able to identify research requirements.

Acceptability

Researchers need to recognize that scientific information may affect vested interests, be they economic, social, or political. Understanding the nature of any vested interests is crucial to maximizing the acceptability of science outputs. Acceptability can also be increased by engaging users in designing and steering the science from the outset. This is easiest if the researchers are based within the user community itself or are, in principle, mandated to advise key actors or the system as a whole. The acceptability of scientific information can be increased if the researchers are actually located within the policy system that requires the information. The receptivity of audience is an important aspect, and the overall acceptability of the research findings will be increased if receptivity is high.

Increasing Acceptability of Science Output

The CRN on disease implications of climate change offers a good example of how the acceptability of science can be enhanced if the research group is part of the policy-making system (Confalonieri, chapter 22, this volume). Most of the Co-PIs from Brazil, Venezuela, Mexico, and Colombia were associated with national public health institutes, which advise ministries of health. Although located in academic centers, these scientists were fully aware of national health agendas and disease concerns. This CRN involved the development of advanced tools, based on new science, for epidemiological surveillance and planning disease control. The users of these tools include public health officers in charge of disease-control activities, who may not accept the value of the underlying science.

The local adoption of these tools can be enhanced by several means. Public health offices can be encouraged to conduct local pilot projects to demonstrate the utility of these tools. Overcoming the risk aversion of public health officials by better explaining the uncertainties involved can be helpful in promoting acceptance. For example, an 80 percent probability of increase/decrease in a disease may not be regarded as enough for a decision maker to risk career or position. Acceptance can also be achieved by helping smaller or poorer municipalities and regional governments to understand the relevance of climate to disease occurrence.

Researchers need to balance the emphasis placed on these different considerations. If, for instance, there is overemphasis on credibility, some key aspects relating to utility might be missed or neglected in discussion with policy makers. The policy process uses all aspects discussed above to some degree, and the trade-offs between different aspects need to be thought through in advance.

Lessons Learned

IAI CRNs and other regional and international research projects provide an array of experiences that shed light on the interface between science and policy. A consistent, emerging lesson is the need for scientists to take the time to understand how the policy process operates and identify its key actors. Scientists need to be aware of the context of their work and be flexible in when to engage with the policy process. They also need to recognize certain inherent characteristics of policy making. Most important, scientists must accept that policy making is an ongoing and iterative process. In particular it is not a linear process whereby research is undertaken and its conclusions are then communicated to policy makers at the end of the research project. Scientists must strive to engage policy makers at various stages of the research process, including and especially during the initial framing of the research questions. These scientists must then be ready to engage in an ongoing communication, repackaging and updating the messages as appropriate. This requirement is unfamiliar to many scientists, who think their role is simply to present their findings only once. When interacting with policy makers, scientists must go beyond conventional publishing formats. While each CRN had its own characteristics, many key generic lessons have emerged from the analysis discussed above. These are presented here with no implied priority.

Establishing and Maintaining Credibility

- Objective and credible science is essential to inform public policy.
- Building and maintaining trust and credibility between all partners is needed from the outset.
- Trust is fragile, and when lost, recovery may be difficult.
- Long-term, stable, and adequate science funding enhances credibility.

- Scientific credibility is ensured by the meeting of international standards for review and the archiving of data and publications.
- Involving some PIs who are both based in the policy development process and potential users is advantageous.
- Scientists who gain the trust of users have an advantage in securing resources for follow-up activities.

Achieving Practicality

- Scientists need to be aware that, having established the practical utility of their work, they may face demands for continuing commitment.
- An opportunistic approach to engaging with the policy process can be effective, but scientists need to be both flexible and able to respond quickly.
- Scientists need to be open to possibilities that may be beyond their initial objectives, to meet new needs as policy evolves.
- Research planning must recognize the social, political, and economic contexts, and the involvement of social scientists is required.

Demonstrating Utility

- In influencing policy, "good science" is not enough.
- The potential relevance of scientific output in the science-policy dialogue may not be immediately recognized.
- Fundamental research and monitoring can have long-term relevance for policy formulation.
- Basic research output is unlikely to fit the needs of the policy process.
- Scientists need to be flexible and open to the possibilities of new users.

Providing Accessibility

- Scientists must recognize the need to communicate with the user either directly or indirectly in understandable terms.
- Developing a common language for communicating between scientists and users is a learning process.
- Making science accessible to users in projects originally designed to produce fundamental knowledge can be achieved by beginning at a local level.
- Explanation of complex science can be done through use of empirical observations as examples.
- Communicating science findings needs appropriate formats that may be novel to science (e.g., videos, pamphlets for practitioners).

Ensuring Acceptability

- Scientists need to establish the way in which users of science output understand uncertainty and deal with natural variability.
- Establishing trust and demonstrating commitment facilitates acceptability.
- Acceptability of science output may be enhanced by mediation through actors who have had a long-term involvement with policy makers.
- Scientific knowledge can be made more accessible by the development of decision support systems that target immediate policy and operational use.
- Pilot studies can demonstrate the usefulness of scientific knowledge and overcome possible reluctance to face the risks of using new, untested tools.
- Publishing in high-visibility journals can attract media attention and thereby influence acceptability of new science by policy makers at multiple levels.

CRNs' Engagement with Policy Process

Experiences from IAI's CRNs have provided several examples of the many different ways that scientists have sought to share their research with policy makers. Some projects specifically sought to have an influence at the local community or individual scale. These research projects intentionally tapped into local traditional knowledge to try to influence policy from the ground up. They chose to attempt to influence human behavior and activity rather than focus on higher-level governmental decision-making processes. In many cases, this approach did ultimately result in higher-level influence.

Projects varied in the timing of their engagement in the policy process. Some projects brought in stakeholders at the mid-research point. In these cases, policy-relevant individuals, such as consultants or advisors, were generally brought in to achieve a more-effective interaction. Many projects also made concerted efforts to "take their findings to the people" at the end of the research phase through outreach activities such as workshops and seminars. These were often targeted specific policy makers or civil society organizations who might influence policy process and its outcome. Some projects, however, chose not to address policy issues at any stage although, ideally, scientists should engage users from the very beginning of research planning.

Conclusions

The IAI enterprise was launched by governments to improve global change science and build capacity in the Americas to contribute to the science-policy enterprise. As an element in this establishment, governments were important in outlining a policy-relevant scientific agenda for the IAI. The scientific community responded with willingness by proposing CRNs to address the IAI agenda. However, many scientists had little experience in bridging the science-policy interface, and proposals typically focused on advancing fundamental science. Nevertheless, in the process, each of the CRNs developed different approaches appropriate to its field of science and user group. This assessment has provided a valuable means of distilling lessons that should assist in the management of the IAI and increase the ability of scientists in the region to contribute to the policy-making process. The assessment has clearly demonstrated that learning how science can best contribute to developing policy is an interactive, ongoing process and that the IAI can make a valuable contribution as it evolves.

Literature Cited

Lara, A., D. Soto, J. Armesto, P. Donoso, C. Wernli, L. Nahuelhual, and F. Squeo, eds. 2003. *Key scientific components for a national policy on the uses, services and conservation of the Chilean native forests.* Valdivia: Universidad Austral de Chile, Iniciativa Científica Milenio de Mideplan. (134 pp.)

5

Communicating Science to the Media, Decision Makers, and the Public

Theo Beckers, Maureen Woodrow, Paul E. Filmer, Sônia M. F. Gianesella, Laura Gallardo Klenner, Carlos A. Klink, Jean-François Tourrand, and Peter Weingart

Communicating interdisciplinary scientific knowledge on global change to political representatives, state agencies, nongovernmental organizations (NGOs), civil society groups, and the public at large is not an easy task. Global change research deals with complex issues of understanding (cognitive complexity), valuing (normative complexity), and acting (policy complexity). Improving communication of complex issues through a common scientific language can improve understanding across different scientific communities. Ecosystems are interrelated with surrounding socioeconomic systems. Linking social and policy science approaches with those of the natural sciences is therefore indispensable. Interdisciplinarity not only gives added value but enriches the research results, as demonstrated in a number of the Inter-American Institute for Global Change Research (IAI) Collaborative Research Network (CRN) projects.

The many actors involved in research on the global society need to improve their problem-solving capacity. Emerging forms of governance include many stakeholders, which make science-policy interactions both difficult and challenging. Sound scientific information can help increase consensus and cooperation and may reduce normative and policy complexity. To link science and policy, global change research must consider societal demands and translate scientific knowledge into action. Communication plays an important role in the relationship between science and society and internally between the sciences. Different modes of communication are practiced within the IAI CRNs. Most important is the traditional communication between scientists from different disciplines and nations, resulting in presentations, publications, teaching, and training. Mutual learning between researchers and others (such as NGOs, decision makers, and those working or living in the area, such as fishers or landowners) occurs when these "others" are directly involved in the area of investigation. Researchers often start by visiting the planned area of investigation to learn from the people who experience global change problems. Researchers also consult decision makers to understand the nature of the policy interventions either underway or anticipated. Both these sources of knowledge are integrated into the scientific information and are important to the research process. Finally, researchers discuss their findings and results with those consulted. The standard model of science communication is based on the principle that science has the monopoly on truth and has the right to define problems and their solutions. A more realistic and effective model of communication might be based on the idea of multiplicity of knowledge. The art to doing research is enriching expert knowledge with knowledge embedded in organizations, institutions, and people. This communication reaches out to the general public, policy makers, and stakeholders who are not connected to the research project.

Communication has a function to clarify, give feedback, and create commitment. Communication not only must contribute to effective policy making but also should improve the research process. It involves two components: sending and receiving. Sending includes a tailor-made, targeted, verbal or visual format. Receiving implies selective interpretation by users. The cultural and structural barriers in the process of packaging, sending, receiving, and interpreting need critical attention. *Effective communication and social learning are always specific and context-bound.* Education and training are essential pillars of the bridge between science and policy. *Practical* examples of communication and outreach mechanisms are *well-designed* newsletters, Web sites, list servers, annual reports, and training institutes. A positive outcome of IAI training institutes is the formation of networks of scientists and professionals.

The effective input of science to policy could be enhanced by taking into account the differences in temporal and spatial scales. The *research* life cycle consists of the following stages: development of an idea and of hypotheses to be tested, establishment of international networks of researchers, proposal development and funding, capacity development, data gathering, analysis, final reporting, and dissemination of results to scientific and nonscientific communities. This has to be synchronized with the *policy* life cycle stages: diagnosis, planning, decision making, and implementation. Systems of science and policy also operate on incongruent spatial levels: global, regional, national, and local. Cross-cultural awareness and output in the language of those most impacted by the research is an important element in transnational communication. Different moments require different types of knowledge and different modes of communication.

Rationales of Science, Politics, and Publics

Scientists are used to thinking that their arguments should be taken on the basis of the evidence provided, that is, methodology, data, and argued conclusion. When they present their arguments to their own community, they assume the responses, even though they may be critical, will be based on the same disciplinary frame of reference. However, in communicating with colleagues from other disciplines, this assumption no longer holds. The problems of other disciplines are different; so are their subject matters, their methods, and their ways to argue a case. The world of scientific knowledge is highly differentiated into many disciplines, subdisciplines, and specialized areas of research, each with its own culture (Forman 1971; Sagar 2000; Dalgaard et al. 2003; Latour 2004; Göbel, chapter 9, this volume). There is an increasing plea for interdisciplinarity to overcome this issue (Stehr and Weingart 2000).

The problem of the different types of communication is exacerbated when we leave the world of science, in which communication is based (at least in principle) on rational argumentation, and move into the world of politics. Here an argument is neither primarily nor exclusively taken on the basis of its rationality. Rather, what counts is the way in which an argument serves to advance or threaten political power (Carrier et al. 2004). In the world of politics, scientific knowledge is used primarily in its strategic function for the preservation and advancement of power (maintaining legitimacy) and secondarily in its problem-solving function (Roqueplo 1995).

Communicating scientific knowledge is also difficult when experts present their arguments to the general public. First of all, they cannot count on their audiences to understand their language unless they find ways to express themselves in nontechnical terms. Because of specialization within the scientific community, popularization has become a profession in itself. In addition, the public may not be interested in the scientists' arguments even if they understand them. Most knowledge produced by scientists has very little apparent relevance to the general public. If it is relevant, it is so for a specific or targeted clientele. Typically, the general public judges the relevance of scientific knowledge on the basis of its immediate value to them, as opinion polls invariably show (Wellcome Trust 2000). Medical information on health risks or information on natural disasters and property risk are high on the list.

Most communication to the public is through the media. In fact, communication scholars and political scientists concur that the media *are* the public (Donges and Imhof 2001). The media are not transmitting knowledge objectively. The communication process changes the information as it is transmitted. Obviously the media select information from the vast amount that is accessible to them, and they do so according to their perception of what will interest their readers or viewers. The media are usually commercial enterprises that have to attract audiences—the larger the better. The criteria according to which they select and present information are "news-values": drama, novelty, human interest (e.g., emotionality), and familiarity (e.g., geographical proximity). Scientific knowledge usually lacks all these attributes and, therefore, is either not taken up by the media at all, or if it is, it will be transformed to fit these values. Scientists often find themselves misrepresented by the media precisely for these reasons. However, given the social and political relevance of global change science, researchers involved in this area can turn these difficulties into an advantage in terms of impact and awareness.

For scientists who are interested in communicating their knowledge to policy makers, the public, and the media in order to influence political decisions and/or behavior, it is crucial to understand how the political system works, how the media operate, what the interests of the public are, and how the public reacts to expert pronouncements. Failure to do so not only leads to frustration on the part of the scientists but may have negative consequences. The scientific community must understand that science is only one element in the policy process—and often not the most important one. Politicians have very different goals from those of scientists. They must mobilize public support for their objectives and convince electorates. Science is a costly enterprise that relies on public support for the resources it requires, at least in democratic societies. This dependency makes it mandatory for scientists to communicate to politicians, to the media, and to the public at large in ways that they understand and that address their concerns. Communicating science and its relevancy to society is now part of the funding process. Science communication has assumed greater importance than ever before (Norton 1998, Bennett et al. 2003, Lubchenco and Mehta 2004, Palmer et al. 2004, Castillo et al. 2005, Nowotny 2005).

Lessons Learned

The policy relevance of each scientific project is not always apparent to the scientists. Popularizing science is not an explicit part of the project, nor is it perceived as important by scientists at the outset. Many scientists do not pay attention to the need to communicate beyond scholarly journals and networks, believing that these communication mechanisms are sufficient. In fact the advancement of scientists within their institutions is based almost exclusively on scholarly communications. Yet showing the relevance of science and of what scientists do was identified by the IAI as an essential element in the funding process, following a review of the first round of research funding in 1997.

The CRN program approved in 1998 included demands for producing information for policy and decision makers. Thus the communication of science to society was made the responsibility of all scientists in the research team. A communication plan or strategy was not requested as part of the research proposals. Analysis of the experience from projects funded in the past 10 years allows us to emphasize the importance of anticipating and understanding communications issues that may be either controversial or advantageous to the research process. Planning how researchers will communicate to those outside the research community, such as stakeholders, media, policy makers, and politicians, from the outset can help scientists avoid pitfalls in the research process. An understanding of the options and how they may be used most effectively for communication in a project is critical from the outset. The levels of communication involved in these complex projects are international, national, regional, and local. Communication occurs among and between scientists who are part of a project, between scientists and policy makers and politicians and NGOs, and also at the community and local level where the research is being carried out. Literacy levels will vary within a project's countries and localities and will require different mechanisms of communications to ensure the message is understood. All projects are collaborative and interdisciplinary, involving at least four countries but possibly as many as seven or eight. Each requires different types and sensitivities of communications. A strategy or plan from the beginning can facilitate the research process. The international nature of all projects involves at least two languages and potentially many more, including indigenous languages. At the national and local levels, the principal investigators (PIs) and co-principal investigators (Co-PIs) have a responsibility to inform the relevant politicians, policy makers, and NGOs of the research funding obtained. The type of information provided or the interest in what is being done by the researchers will vary depending on the type and nature of the project. Effectively communicating the project, its goals, and its objectives to the public is likely to have more impact than a scientifically oriented explanation. Not all projects will garner the same degree of interest in their early stages. Those that focus on filling data gaps will have different requirements for communications than those working with the human dimensions of science. Research with an obvious human element, such as health projects or those dealing with natural disasters, may be of greater interest to policy makers and politicians than those involving a comparative analysis of ecosystems. Yet all projects have the potential to be controversial at a local or national level and have downstream effects on current and future research in the region. Local communities have long memories, and a naive scientist may be in that memory.

Communicating to policy makers requires special skills. CRN projects are carried out across disciplines and international borders. Not only are there barriers between policy makers and researchers within countries, but the international dimension of these projects requires understanding of the policy dialogue and its differences in the various countries. A Brazilian Co-PI of a CRN project on ultraviolet (UV) radiation effects in natural ecosystems, who spent a month with her Canadian counterpart, was surprised at the level of consciousness and interest in UV exposure in Canada. In her country, the press does not have the same interest, in spite of the high natural UV levels in Brazil. In contrast, in Canada the interest in UV exposure is high. Many governments have increased awareness of UV radiation and exposure to the sun by including UV exposure risk in the daily summer weather forecast. Thus sensitivity to policy differences in the countries involved can be a useful tool in comparative analysis.

How can scientists communicate to the public in a manner that not only reaches but is understood by the general public? Do mechanisms exist within the research institutions, the scientific community, or the funding agencies to assist scientists with communicating their science to the public? Scientists are not trained to interact with the media. Failure to recognize the importance of this communication can slow down or impede public understanding. Some scientists are politically savvy and a have personal talent with communication, and this talent can enhance their research. However, it can also place a burden on a talented communicator who is part of a larger group of scientists.

The research process, as outlined earlier, includes a number of phases: developing ideas and establishing a hypothesis to be tested, establishing international networks of researchers, proposal development and funding, capacity development, data gathering, analysis and synthesis, preparation of final reports, and diffusion of the results to the scientific and nonscientific communities. During these various phases, different communications skills are necessary. Communications during the first three phases have traditionally been within the scientific community; however, the planning phase can involve meeting with stakeholders and policy makers to capture their knowledge and encourage their participation in the development of the research hypothesis (Norton 1998; Castillo et al. 2005). If research is to be truly policy relevant and useful for decision makers, integrating them into the research process from the very beginning will ensure the research questions formulated are related to decision makers' needs. Of the 13 CRNs funded in 1998, approximately 70 percent received prior seed grants to develop their proposals. The funding was used to hold meetings not only with scientists to build the research network but also with potential stakeholders who would participate and have an interest in the research outcomes. The Andean Amazon rivers analysis and management project used the seed money for a workshop that included not only the scientific community but also policy makers, NGOs, and potential stakeholders of the project (McClain et al., chapter 19, this volume). When the scientists obtained funding from the IAI, the experiences from the planning workshop led researchers to include the appropriate participants at the community level.

Capacity building refers mainly to that within the scientific community, but it can also occur at the regional or community level through workshops or training sessions or community education. An added plus of good communication on the nature, design, and significance of research can be encouraging the participation of good graduate students and researchers. Good Web sites organized and structured in a clear manner are excellent means of communicating to not only graduate students but others interested in the research. Training graduate student researchers to communicate in the field is critical. A project on cattle ranchers and deforestation required extensive training of field-workers to ensure that the researchers were aware of local and national conflicts on the issue (Wood and Tourrand, chapter 17, this volume).

During the field-research and data-gathering periods, communication not only in the right language but also to the right people is a vital element of success of the project. A project on climate variability and its impacts in Mesoamerica and the Caribbean produced two documentaries that led not only to several TV and press interviews but also to training programs in high schools, teaching students and teachers how to use the meteorological information (Magaña and Caetano, chapter 13, this volume). The press exposure led to a positive reception of research in the field. The analysis and synthesis is mainly a scientific endeavor, but communicating elements of the results back to the community and to policy makers can make a difference in the utilization of the scientific results. While preparation of the final report is purely scientific, the production and diffusion of both an executive summary and briefing notes on key elements of the research are essential for practitioners and decision makers. Scientists also have a responsibility to disseminate scientific information and engage global scientists in a dialogue with policy makers, government agencies, and the private sector. Journals such as *Science* and *Nature* are excellent for this scientific debate. However, to encourage the scientific debate on a global level, the major journals are in the English language. Similar broadly based scientific journals are not available in other languages.

Generally, policy makers have shorter time horizons than scientists, and results are needed immediately. Even if the needed research findings are available, the language in which they are expressed is too complex. To effectively communicate, scientists need to understand the policy process and provide the information required by the policy makers in a concise and relevant manner. Involving them at an early stage, not only as part of the research process but also in the peer review of the research to be funded, can be useful in ensuring that the research proposed and carried out is presented in an effective manner.

The task of communicating effectively requires a team effort of not only scientists and policy makers but also communications specialists and possibly editors or members of a production staff. Nurturing the media and policy communities, providing them with regular information and data of importance along with stories of human interest, can promote better communications of the research within society in general.

Recommendations

- Develop a communication strategy at the organizational level as well as at the project level.
- Develop news outlets that communicate the importance of research to policy makers and interested publics.
- Develop connectivity strategies between regional findings on global environmental change and international agendas.
- Provide guidance in the development of outreach elements on project Web sites for the different types of funded research, and support the management of the sites.
- Enhance training of all researchers, especially for newly funded projects, in the skills required for communication with the media, policy makers, politicians, NGOs, business, and the interested publics. This training should include the value of communicating beyond traditional science.
- Require a section in the research proposals on objectives and strategies of project communications.
- Improve the documentation of research that successfully integrates the concerns of decision makers and stakeholders. Good examples will help researchers and those involved in the research process in the translation from science into policy.
- Since funding agencies increasingly request policy relevance and communication to policy and society, as part of their funding criteria, funding agencies in turn should

give recognition to external communication and communication skills within the research and academic institutions.

Literature Cited

Bennett, E. M., S. R. Carpenter, G. D. Peterson, G. S. Cumming, M. Zurek, and P. Pingali. 2003. Why global scenarios need ecology. *Frontiers in Ecology and the Environment* 1(6):322–9.

Carrier, M., D. Howard, and P. Weingart. 2004. *Science and values.* Notre Dame, Ind.: University Press of Notre Dame.

Castillo, A., A. Torres, A. Velázquez, and G. Bocco. 2005. The use of ecological science by rural producers: A case study in Mexico. *Ecological Applications* 15(2):745–56.

Dalgaard, T., N. J. Hutchings, and J. R. Porter. 2003. Agroecology, scaling and interdisciplinarity. *Agriculture, Ecosystems and Environment* 100:39–51.

Donges, P., and K. Imhof. 2001. Oeffentlichkeit im Wandel. In Jarren, O., and H. Bonfadelli, eds. *Einfuehrung in die Publizistikwissenschaft.* Bern: Haupt.

Forman, P. 1971. Weimar culture, causality and quantum theory, 1918–1927: Adaptation by German physicists and mathematicians to a hostile intellectual environment. In McCormmach, R., ed. *Historical studies in the physical sciences,* vol. 3. Philadelphia: University of Pennsylvania Press.

Latour, B. 2004. *Politics of nature: How to bring the sciences into democracy.* Cambridge, Mass.: Harvard University Press.

Lubchenco, J., and G. Mehta. 2004. International science meetings. *Science* 305:1531.

Norton, B. G. 1998. Improving ecological communication: The role of ecologists in environmental policy formation. *Ecological Applications* 8(2):350–64.

Nowotny, H. 2005. High- and low-cost realities for science and society. *Science* 308:1117–8.

Palmer, M., E. Bernhardt, E. Chornesky, S. Collins, A. Dobson, C. Duke, B. Gold, R. Jacobson, S. Kingsland, R. Kranz, M. Mappin, M. L. Martinez, F. Micheli, J. Morse, M. Pace, M. Pascual, S. Palumbi, O. J. Reichman, A. Simons, A. Townsend, and M. Turner. 2004. Ecology for a crowded planet. *Science* 304:1251–2.

Roqueplo, P. 1995. Scientific expertise among political powers, administration and public opinion. *Science and Public Policy* 22(3):175–87.

Sagar, A. D. 2000. Capacity development for the environment: A view for the South, a view for the North. *Annual Review of Energy and the Environment* 25:377–439.

Stehr, N., and P. Weingart. 2000. *Practising interdisciplinarity.* Toronto: Toronto University Press.

Wellcome Trust. 2000. *Planning for the future: The Wellcome Trust 2000–2005.* London: Wellcome Trust.

PART TWO
Background Chapters

6

Communicating Science in Democratic Media Societies

Peter Weingart

Outdated but Persisting Models of Science Communication

A gap exists between, on one hand, scientists, science administrators, and the professionals working in "public relations" for science (such as agencies organizing science events or producers of science magazines and journalists) and, on the other hand, specialists in science and media studies doing research on science communication. The former think of themselves as expert practitioners, and the latter think of themselves as knowing better because they have actually done research on the practitioners' performance. The former hold on to what the latter call an outmoded "canonical model." This model informs most of the public understanding of science (PUS) and public engagement with science and technology (PEST) activities and many science journalists in their work. It is outmoded because of its basic assumptions about the communication process, about the publics that are to be addressed and why they should be addressed, and about the relation of science to the media and to the general public (Shinn and Whitley 1985; Hilgartner 1990). What are these assumptions, and why are they misplaced?

The first assumption is that the public suffers from a deficit of understanding science, and this lack of understanding explains the public's resistance to new knowledge and new technologies derived from science. Conversely, if the public was better informed and more knowledgeable about science, it would be more positively inclined toward scientific progress and its products.

The second assumption is that, as scientific knowledge is much too complicated to be understood by the average citizen, it needs to be mediated. This mediation must meet certain conditions: It must attempt to capture the public's interest, which can only be attained by using the means employed by mass media (edutainment, infotainment), and at the same time, it must be true to the contents of the information disseminated by the scientists. The many forms of mediation come under the label *popularization* and are referred to science journalists, professional popularizers, organizers of exhibitions, and others—all of whom may be summarily called "the media."

The third assumption is that there is a clear distinction between science (i.e., individual scientists or science organizations such as universities, research councils, science journals) and the media as defined above. Science, in this understanding, produces new and authoritative knowledge and is politically detached. It disseminates this knowledge without any interest attached other than pronouncing the truth. The media, on the other hand, are relaying this knowledge to the public in an understandable way, but they suffer from lack of understanding due to their generalist competencies. They tend to apply inadequate criteria of selection and representation, instead focusing on the sensational, emotional, irrational, and so on. The underlying model of communication is one of "unidirectional, linear transfer from one sender (the scientific community) to a passive receiver (the broad uninformed public)" with the media achieving the transfer and being responsible (and consequently being blamed) for "distortions" (Bucchi 1996, p. 377).

Part of the answer to why these assumptions are misplaced lies in explaining how they came about in the first place, that is, their historical origins and their social and institutional justifications.

One has to realize that science as a social institution has changed over time since its modern inception in the 17th century. With it the political environment of science, that is, the structure of societies and the organization of political power (governance) supporting science, as well as the publics being addressed by science, have fundamentally changed over the last three centuries (Fox and Turner 1976; Biagioli 1993; Guston and Kenniston 1994; Daum 1998). This change may be represented by the development in science patronage:

- aristocracy—science patronage by the courts (17th–18th centuries)
- educated bourgeoisie—science support through philanthropy (19th and early 20th centuries)
- mass democratic electorate—science support by popular vote (second half of 20th century–present)

The type of patronage implies the ways and means by which science addresses its patrons. But it is necessary to understand why science is depending on patronage in the first place. The reasons lie in the differentiation of science as an independent social system. On the level of institutions, the main development is the emergence of disciplines in the early 19th century. It implies the self-direction of research by the internalization of the criteria of relevance and of quality standards. This occurs parallel to the establishment of extended training and the state accreditation of degrees. On the level of contents, this implies an increasing abstraction of subject matters, the creation of a laboratory world, the mathematization of scientific language, and the emergence of specialized languages and roles of experts (Stichweh 1984).

All these developments contribute to the impression that science is more and more distant from society and inaccessible for the average citizen. In fact the differentiation of science is part of the functional differentiation of societies in general. But this process creates dependency. Science does not produce the means for its own subsistence. Even today it does so only to a small degree. Science needs support from society, which means that it becomes one item in the overall state budget; it takes part in the general political process of resource distribution. The agents who determine this process, from absolute rulers to democratically elected policy makers and their electorates, are therefore the obvious addressees of appeals from science to decide in its favor.

Given an increasing distance between science and the rest of society, and at the same time given science's dependency on society's resources, it is obvious that the task of mediation assumes importance. It should also be apparent that this mediation takes on different forms depending on both the target audience and the information available to the scientists. With some simplification, the types of mediation can be associated with types of audience, following three historical stages:

- In court societies, communication is from "below." Science seeks interest for curiosity and public experimentation; curiosity cabinets exist as entertainment.
- In bourgeois societies, communication is from "above." Science has the monopoly on truth and the definition of problems and solutions and is popularized in popular journals and amateur naturalist associations.
- In mass democracies, expansion of demands for participation requires new types of communication. Science is reported by the mass media (print and TV) and staged in public events (PUS, PEST, science dialogue) or in science centers (Weingart 2005, ch.1).

The canonical model of science communication rests on a new image of the public that emerged in the 1920s and 1930s. The knowledge-hungry public of the 19th century that read popularizations of science in journals and attended public lectures was replaced by the new mass public that, from the perspective of the scientific community, appeared disinterested and incompetent to understand new scientific developments. The reasons for this shift are most likely to be found in a widening gap between the increasingly influential lower classes and the dwindling bourgeois middle classes (from whose ranks most scientists came) that led (in Europe) to a general mood of cultural decay and crisis (Forman 1971). Contemporary advances in physics, relativity theory, and quantum mechanics added new complexity and inaccessibility to scientific knowledge while assuming world view function at the same time (Bensaude-Vincent 2001, p. 109). Popularization lost its function and became looked upon by scientists as at best superfluous and as "contaminating" scientific knowledge (Hilgartner 1990). After World War II this development continued as a result of the demise of the bourgeois elites and the democratization of the industrialized societies. It is perhaps indicative that when the first nuclear accident occurred at Three Mile Island in 1979, nuclear engineers, faced with public protests, declared the fears of protesters to be irrational and unfounded. This was a turning point, however, in the denigration of the public and the media by the scientific community.

The appearance of scientists on TV, advocating nuclear energy or its demise, that is, taking a political stand in a public debate, destroyed the image of a clear separation of apolitical experts and political decision makers. This has led analysts to postulate a continuum of science communication rather than a sharp distinction (Cloitre and Shinn 1985; Hilgartner 1990).

New Challenges to Science Communication

The antinuclear debate of the 1970s and 1980s was triggered by the high-handed implementation of a nuclear technology at a time when the protection of the environment had emerged as a competing value that rapidly spread through all industrialized countries. The political, scientific, and corporate elites misjudged the impact of this new value set, which translated first into new social movements and soon into more stable nongovernmental organizations such as Greenpeace and political parties like the green parties—all indications of a new wave of democratization. Much of this process was focused on the perception of risks, both to the environment and to human health, associated with new technologies (Beck 1986). Nuclear energy had become the paradigm case of scientific/engineering knowledge being implemented with various degrees of uncertainty and lack of knowledge of its impact. Media reporting revealed dissent among experts and thereby uncovered the hitherto hidden process of internal scientific debate to the public at large. Technology and science policy were given a new reflexive reference in that they had to be oriented to the protection of human health and the environment (i.e., *technology assessment* or *risk assessment*, which later was to be summarily labeled *sustainability*) (Weingart 2004).

A second major challenge to science communication arose with the changing role of the media. Beginning in the interwar years and continuing at a faster pace after World War II, the media separated from the political system. Publisher families and political parties as owners of print media declined in number and were replaced by large corporations whose primary goal was financial profit, thus addressing a mass public as customers. This differentiation of the media as an independent social system implied that they created their own criteria for selection and evaluation of information that no longer followed the principle of "true representation" (Donges and Imhof 2001, p. 121).

Insofar as science provided news items, they became the object of a new brand of professional science journalists who operated under the constraints of the media as economic enterprises replacing the popularizers. The content and form of their reporting were dictated by the news values, which were to attract the attention of as large a public as could possibly be reached. This explains why science reporting focuses on scientific controversies, on dramatic scenarios such as floods or droughts associated with climate change or meteorites hitting Earth, on failures of scientific predictions, and on uncertainty of scientific knowledge, that is, on risks, on challenges to ethical values such as cloning, and on moral issues such as fraud in science. Enlightening the public is only of secondary importance. Even the exceptional examples of first-rate science journalism in leading newspapers and quality-conscious television do not completely escape the exigencies of orientation to their imagined publics. It is naive to expect the media to mirror the scientists' perception of what is newsworthy about their work and the knowledge they communicate to their peers.

A third challenge is the fact that science has lost its status as an estate, that is, as an institution whose use of public funds and internal quality controls have largely been taken for granted on the promise of the ultimate value of its findings for society (Price 1967). The impact of the "audit culture" has not stopped at the walls of the ivory tower. Since the end of the Cold War, the research community has come under public observation of its internal mechanisms of priority setting and attribution of reputation. To some extent, those mechanisms have become externalized under the mandate of a demand for accountability (Guston and Kenniston 1994; Power 1997). Scientists across the globe are under pressure to publish their work in ways accessible and visible beyond the narrow circle of their colleagues and at a pace that makes them eligible for the remunerations that are now attached to publication and citation records. The temptation to rush into the media to secure priority is particularly strong in fields in which the financial stakes are high and the half-life of novelty is short: medicine, biochemistry, materials sciences. This temptation is enhanced when the media are very responsive and readily disseminate information given to them because it meets the usual news values. Such information may then have, due to its effective and widespread proliferation, an enormous impact on public discourse and, ultimately, on decision makers in governments and in the corporate world. The danger of initiating mass reactions, and thereby triggering action on the part of those affected directly or indirectly, has grown as the entire media system observing society has come to be highly sensitive. The tight coupling between science, the media, and politics exacerbates the particular risks of communication faced by each.

The fact that science, politics, and the media all operate with their own criteria of relevance means that the perception and communication among them are highly selective. Information given by scientists to the media may be amplified beyond recognition, or it may be muffled or altogether ignored. The profit motive will induce the media to use the information in any way they deem fit. Policy makers may overreact, or they may process the information in contexts that may have very little to do with the message as it was intended by its senders. Their fear of losing legitimacy will motivate policy makers also to try to control the appearance of knowledge. Finally, if scientists engage in advocacy or if they try to gain attention by communicating exaggerated warnings of impending catastrophes, they may lose their credibility. Thus the different rationales of science, politics, and the media create different risks of communication specific to each:

- for science, loss of credibility
- for politics, loss of legitimacy
- for the media, a loss of attention and market shares

These challenges taken all together render the assumptions that still govern much of the understanding of the relationship between science, on one hand, and the general public, the media, and politics, on the other, untenable. It is crucial that this much more complex model of communication replace the simplistic, unidirectional one that still informs the discussions about science and the public.

Science Communication and Its Dangers: The Case of Climate Change

Climate change research is a particularly pertinent case because it involves scientific knowledge that is policy relevant, and it is of interest to the media, not least because it was communicated by scientists who advanced sensational warnings of an impending catastrophe. (It is not a case, however, that exemplifies *all* aspects of science communication mentioned above). The focus of the following is on the communication of climate change in Germany, simply because it is a case that has been studied thoroughly and the data are readily accessible, but the patterns observed repeat themselves in other countries (Weingart et al. 2000).

The publication of a short report by German physicists declaring that climate change would have a dramatic impact on sea level was motivated by a desire to put nuclear power back on the agenda as an option of energy policy. The declaration triggered a long-lasting debate and ultimately led to the launching of an extensive research program on climate change, although the initial claims had to be corrected later (Weingart et al. 2000). The intended effect of the dramatized reports on climate change—to get nuclear energy back on the political agenda—was not achieved. In the German parliament, the threat was taken seriously. It led to the establishment of a national climate research capacity and, ultimately, to the commitment to a 25 percent reduction in carbon dioxide emissions. This goal, however, could not be achieved and threatened the government's legitimacy. Between 1975 and 2000, the issue of climate change was linked to different issues on the political agenda. That scientific knowledge will be used by politicians according to their interest is demonstrated by two parliamentary committees on climate change. While one managed to achieve consensus across political party lines, the second is generally considered a failure because the consensus was lost as politicians positioned themselves for upcoming elections.

The media reacted predictably by taking up the dramatic warning of the sea level rise. This was exemplified by the cover page of a German news magazine (*Der Spiegel*) showing the Cologne Cathedral half submerged. Since climate change is not part of everyday experience, the media typically translate the abstract phenomenon into experiential categories according to news value. Unusually warm summers, heavy rainfall, and spring floods are associated with climate change, as is the lack of snow for winter sports.

Perhaps most striking was a growing skepticism among the media, albeit in a minority of the reports recorded, indicating the usual short-term shifts of media attention and the need to generate news. Skeptical articles questioned the reliability of computer climate models, highlighting the uncertainty of climate change predictions and articulating suspicion that the scientists had overstated their case in order to obtain funds. The credibility originally attributed to the climate change scientists was thus withdrawn, threatening broader public support for their research.

Conclusions

The climate change case illustrates the following:

- The assumed separation between science and its publics, as well as the media and politics, does not exist. Scientists communicate actively and with their own motives.
- At the same time, science is under constant observation by the media, which report in ways governed by their own operations and independently from the intentions of scientists.
- The information disseminated by the media can have broad impact beyond the control of the scientific community.
- The same applies to the processing of the scientists' knowledge by policy makers.

The problem, then, is how science communication can be improved. Since the structural conditions that account for the situation cannot be changed, one cannot expect remedies but only gradual shifts. Most important in this respect appears to be that scientists try to regain a measure of institutional distance from the media and from politics. This means that they should be more careful and sophisticated in their dealings with the media and that they should weigh carefully the long-term advantages of credibility against the short-term advantages of media attention. A careful, thoughtful approach to communication of warnings is called for.

Literature Cited

Beck, U. 1986. *Die Risikogesellschaft*. Frankfurt: Suhrkamp Verlag.

Bensaude-Vincent, B. 2001. A genealogy of the increasing gap between science and the public. *Public Understanding of Science* 10:99–113.

Biagioli, M. 1993. *Galileo, courtier: The practice of science in the culture of absolutism*. Chicago: University of Chicago Press.

Bucchi, M. 1996. When scientists turn to the public: Alternative routes in science communication. *Public Understanding of Science* 5:375–94.

Cloitre, M., and T. Shinn. 1985. Expository practice: Social, cognitive and epistemological linkage. In Shinn, T. and R. Whitley, eds. *Expository science: Forms and functions of popularisation.* Yearbook Sociology of the Sciences IX. Dordrecht: Reidel. (pp. 31–60)

Daum, A.W. 1998. *Wissenschaftspopularisierung im 19. Jahrhundert.* München: R. Oldenbourg Verlag.

Donges, P., and K. Imhof. 2001. Öffentlichkeit im Wandel. In Jarren, O. and H. Bonfadelli, eds. *Einführung in die Publizistikwissenschaft.* Bern: Haupt. (pp. 101–33)

Forman, P. 1971. Weimar culture, causality, and quantum theory, 1918–1927: Adaptation by German physicists and mathematicians to a hostile intellectual environment. *Historical Studies of the Physical Science,* III:1–115.

Fox, R., and G. L. Turner. 1976. *The patronage of science in the nineteenth century.* Leyden: Nordhoff.

Guston, D. H., and K. Kenniston, eds. 1994. *The fragile contract.* Cambridge, Mass.: MIT Press.

Hilgartner, S. 1990. The dominant view of popularization: Conceptual problems, political uses. *Social Studies of Science* 20:519–39.

Power, M. 1997. *The audit society: Rituals of verification.* Oxford: Oxford University Press.

Price, D. K. 1967. *The scientific estate.* Cambridge, Mass.: The Belknap Press of Harvard University Press.

Shinn, T., and R. Whitley, eds. 1985. *Expository science: Forms and functions of popularisation.* Yearbook Sociology of the Sciences IX. Dordrecht: Reidel.

Stichweh, R. 1984. *Zur Entstehung des modernen Systems wissenschaftlicher Disziplinen. Physik in Deutschland 1740–1890.* Frankfurt: Suhrkamp.

Weingart, P. 2004. Science in a political environment: The dynamics of research, risk perception and policy making. *EMBO Reports* 5:52–5.

Weingart, P. 2005. *Die Wissenschaft der Öffentlichkeit: Essays zum Verhältnis von Wissenschaft, Medien und Öffentlichkeit.* Weilerswist: Velbrück Wissenschaft.

Weingart, P., A. Engels, and P. Pansegrau. 2000. Risks of communication: Discourses on climate change in science, politics, and the mass media. *Public Understanding of Science* 9:1–23.

7

Institutions as Initiators and Users of Science

Peter R. Jutro

This chapter explores aspects of the relationship between the conduct of environmental science and institutions that support and use it. The underlying premise is that as environmental science has evolved, inquiry into its complexity requires a level of operational integration that is generally beyond the capability of individual scientists, and often beyond single disciplines. Although current science is built upon the work of individual researchers, and traditional ecological knowledge has generally survived without conventional institutional support, effectively addressing the complexity and scale of current major environmental problems requires resources that strain the capabilities of historic supporters of environmental science. Understanding the context within which science is undertaken and used can be critical to its effectiveness. The socio-scientific issues that face the Inter-American Institute for Global Change Research (IAI) are examples of such intrinsically complex problems.

Historical Context for Science-Institution Interdependence

We may mythologize the image of the lone scientist, but in fact, institutional support for environmental science is ancient—easily traceable to Aristotle as a student of Plato's Academy and then as the founder of the Lyceum with the support of his student Alexander the Great. He used his institutional base to research and publish both *Meteorologica* and *Zoology*, perhaps the first examples of observational environmental science (Aristotle 1952, Aristotle 1965, Aristotle 1970, Aristotle 1991). Institutional affiliation ensured an intellectual legacy. In its absence, in the days before scientific publishing and journalism, knowledge would likely not have survived.

Subsequent environmental science was largely supported by institutional patrons. Taxonomy evolved through Linnaeus, whose 1735 work, *Systema Naturae,* was supported by the Uppsala University in the early 1700s (Linnaeus 1735). Although supported by a university, there was a religious basis to the work: The notion was that the study of nature would reveal the divine order of God's creation and that it was the responsibility of the naturalist to construct a classification system that would reveal this order. Linnaeus is considered to be the founder of the binomial system of nomenclature and the originator of modern systematics. As science matured, a transformation of environmental science from observation and cataloguing (by naturalists) to explanation (by scientists) took place, and the motivation for scientific support continued to evolve. The church became a major supporter of environmental and natural science research, in part because its purpose was the glorification of God. William Paley suggested that God's nature could be understood by analysis of what he had created (Paley 1802). Charles Darwin, although wealthy, was initially placed on HMS *Beagle* in 1836 through a university, Christ's College, in Cambridge, and the expedition was supported by the British navy. Perhaps the last influential institutionally independent naturalist was Alexander von Humboldt (1769–1859), whose work provided the model for Darwin's own journal (von Humboldt 1815). It is worth noting that several of the seminal philosophical works that motivated the development of conservation as understood in the biodiversity context arose more as literature than as science and that these were the works of independent scholars. Examples are the works of George Perkins Marsh (1801–1882), Henry David Thoreau (1817–1862), and Aldo Leopold (1887–1948). This serves to remind us that biological diversity issues may have great moral, ethical, and religious components that must be balanced with scientific analysis and argument.

Institutional support has continued as the enterprise of science has become more complex and expensive and as the opportunities for the unattached naturalist have largely faded. From the academies of the ancients through the universities of today, from the church to charitable foundations and museums, from royalty to royal societies and national academies, from governments to intergovernmental organizations, and even from wealthy patrons to private industry, the search of scientists for institutions to support their work has continued.

Even in publishing, the medium of transformation from the conduct to the use of science, the transition has been from word of mouth to monographs to institutionally supported journals and, most recently, to international cooperative information-sharing mechanisms such as the Convention on Biological Diversity (CBD) Clearing House Mechanism, designed to broker information exchange related to the concerns of the CBD. Other examples are NatureServe (http://www.natureserve.org); InfoNatura, NatureServe's Latin American biotic database (http://www.natureserve.org/infonatura); BirdLife International, a global alliance of conservation organizations (http://www.birdlife.org); and DarwinNet, a biodiversity information mechanism sponsored by BirdLife International in Peru and Ecuador (http://www.darwinnet.org). The last two are partially supported by the United Kingdom's Darwin Initiative (http://www.darwin.gov.uk).

In large-scale environmental research, the need for costly technology has driven the trend. Large-scale environmental computer modeling was originally largely supported by the governments of the United States and the United Kingdom. Availability of satellite imagery to scientists was for decades largely supported by the United States, but it is now also available from Canada, India, Russia, Brazil, Israel, the European Space Agency, Japan, and France. To date, only a national government or a large business enterprise has been able to afford to support the research, development, and operational costs of these tools, independent of their research uses. Industrial government alliances are also becoming more common, making the consistent characterization of a research support taxonomy more difficult. An example is geneticist J. Craig Venter's *Sorcerer II* Expedition: an oceanic voyage in search of unknown life forms, the building blocks of which may be usable for solving environmental problems. This is a private effort heavily supported by governmental investment. Another example might be the technical and financial support provided to scientists for the use of geographic information systems (GIS) by ESRI, the company that researches, develops, and markets GIS software. These can be seen as examples of attempting to "do well by doing good."

Socio-Scientific Issues of the Last Decades

The complexity of the environmental scientific enterprise began with the publication of *Silent Spring* (Carson 1962). It grew as the environmental movement of the late 1960s led to the UN Conference on the Human Environment (Stockholm, Sweden, 1972) and two decades later, as problems were conceived of as being increasingly complex, to the UN Conference on Environment and Development (UNCED) in Rio de Janeiro, Brazil, in 1992. Whereas difficult environmental pollution problems were on the Stockholm agenda, they were straightforward by current standards. By the time of UNCED, only stratospheric ozone depletion had been effectively addressed. Social and scientific questions dealing with sustainable development, global climate change, sea level rise, deforestation, desertification, persistent organic pollutants, biological diversity, biosafety, intellectual property protection, and north-south equity all were awaiting some form of approach or agreement. An extensive literature on the history of the environmental movement has developed. Rather than cite an individual text, I refer the reader to a central Internet source on environmental history at http://www.erica.demon.co.uk/EH/EHsite.html. The journal *Environmental History*, which is published by the American Society for Environmental History and contains articles about Latin American environment, is available online at http://www.aseh.net/publications/environmental-history/.

The fundamental nature of these environmental issues characterizes the importance of institutions. Issues have become so complex, multidisciplinary, transnational, and adversarial that it is no longer possible for a single scientist or loose association of scientists to deal with them effectively. Institutional arrangements became necessary, not only to fund scientists, but to facilitate the characterization of complex socio-scientific prob-

Table 7.1. Analytical metrics for environmental problem comparison

	Goal Clarity	Social Complexity	Scientific Complexity
Stratospheric Ozone	High	Low	Medium
Global Climate Change	High	High	High
Biodiversity	Medium	High	Variable

lems, divide them into manageable portions, provide complex technical information, and offer a structure within which such environmental research could be undertaken.

It is also important to acknowledge that a certain identity confusion is historically associated with environmental science. The personal ethic of many environmental researchers is based upon a concern for the environment and a desire for environmental protection. It is institutions that are capable of protecting the scientific integrity of the scientists and their work while collaborating to help ensure that the information produced by the scientists' research helps fulfill their social goals.

Those issues generally characterized as "global" are perhaps the most interesting in terms of the relationship of their scientific practitioners to various institutions. The global environmental problems that were the major concern of UNCED were global climate change and biological diversity, the problem of stratospheric ozone depletion having been addressed earlier by separate international legal instruments. Let us examine what might be required in order to successfully understand these issues and how institutions have evolved to deal with them.

Relative Complexity of Three Iconic Environmental Issues

Both selection of metrics and qualitative characterization of environmental issues (Table 7.1) are highly subjective. Different readers could easily justify the assignment of different measures and values. The table's main purpose is to illustrate differences in essential character of the three problems, in addition to differences in the extent to which science can be effectively applied to their solution. This should suggest the need for further analysis of the optimal nature of institutions that might be brought to bear to influence both problem space and problem solutions for large-scale environmental issues.

Stratospheric ozone depletion, which we now think of in the context of ozone-depleting anthropogenic chemicals, was first postulated to be a consequence of commercial supersonic flight (for documentation on ozone depletion and the issues surrounding it, see http://www.epa.gov/ozone). The chemistry was worked out by three

Scope	Relative Cost	Scientific Uncertainty	Solution Complexity
Global	Low	Low	Low
Global	High	High	High
Regional/ Worldwide	Variable	Variable	High

academic scientists, one Mexican, one Dutch, and one from the United States (Paul J. Crutzen, Mario J. Molina, and F. Sherwood Rowland shared the 1995 Nobel Prize in Chemistry "for their work in atmospheric chemistry, particularly concerning the formation and decomposition of ozone"). The chemistry, although surprising, was remarkably simple. A solution appeared to be relatively simple as well. A limited number of companies manufactured ozone-depleting substances, economic forces pushed them toward the development of more benign substitutes, and supersonic flight did not appear to be economically viable. Treaty negotiations were undertaken to limit ozone depletion under the auspices of the United Nations. Individual governments were urged to recognize that the benefits of compliance far outweighed the costs, and some chemical companies saw economic opportunity in developing benign substitutes. The 1985 Vienna Convention for the Protection of the Ozone Layer entered into force on September 22, 1988, followed by the 1987 Montreal Protocol on Substances that Deplete the Ozone Layer, which entered into force on August 10, 1992. Since the ban, despite episodic masking by annual variability, the concentration of stratospheric ozone has been increasing, and the "ozone hole" has begun to shrink (WMO/UNEP 2006).

Global climate change (GCC) turned out to be a far more intractable problem with a far more complicated solution space. The GCC issue was embroiled in controversy from its beginning for a host of reasons, the largest being the potential economic and distributional impact of attempting to limit it. The major greenhouse gas, carbon dioxide (CO_2), is the product of all combustion and therefore key to the world's energy and economy. A host of distributional issues dealing with, among others, relative per capita energy use and photosynthesis as a countervailing force only complicated the effort of scientists to truly determine what component of global warming could be attributed to anthropogenic forces. Dealing with the question, What side are you on? became a problem as UN treaty negotiations began, generally related to the nature of scientists' identification with a given supporting scientific, environmental, governmental, or business institution. Presumptions of scientific objectivity were broadly challenged, often for political purposes. In some countries, especially the United States, the nature of global change science remains an ongoing domestic political issue.

Biological Diversity

As difficult as those two issues are, perhaps the most complex and most interesting environmental issue, and the one that has the closest ties to Latin America, is that of biological diversity. First of all, the issues' categorizations differ; whereas both ozone depletion and climate change are truly global issues, biological diversity is a worldwide issue rather than a global one. Emissions of ozone-depleting chemicals or CO_2 in one part of the world will affect either climate or stratospheric ozone concentration (or perhaps both) in another. Extirpation of a species or degradation of a habitat in, for example, a Latin American country will not have a direct effect on the biota of, for example, a country in Asia. Analogous problems are distributed around the world, but except to the extent that they are causally related to a truly global phenomenon such as climate change, the likelihood of significant first-order biological effect at a distance is low.

Second, since biological diversity is a concept that is so broadly defined that it encompasses, from a biological perspective, diversity at the genetic, species, community, and ecosystem level, the concept created a vast community of interest among virtually the entire biological science, conservation, and conventional resource extraction communities. Biological diversity will be the focus of the remainder of the chapter, as further illustration of a variety of mutually beneficial science-institutional relationships.

Biodiversity, which arose as an issue concerned largely with protected areas and clearly related conservation issues, emerged from treaty negotiations with a complex assemblage of goals. The CBD addressed not only a broad range of conservation issues but also additional issues such as resource ownership, intellectual property protection, indigenous rights, biosafety, and equitable financial allocation, each of which, even if treated independently, would be challengingly complex.

Types of Institutions

The institution that dominates any discussion of biological diversity is the United Nations (UN). The UN managed the development of the CBD with its secretariat in Montreal, hosted UNCED, maintains the UNCED Secretariat, and supports SBSTTA, its open-ended intergovernmental scientific advisory body known as the Subsidiary Body on Scientific, Technical and Technological Advice. As the CBD was being negotiated, a host of institutions involved themselves in attempting to influence its form. As an international environmental agreement being negotiated in the world of multilateral diplomacy, it attracted a host of governmental units; international, regional, national, and local environmental organizations; industries and industry groups; and think tanks, which all worked to exert influence on the outcome. Many sought to cite scientific information supporting various points of view or to support the conduct of science to do so. This information was directed toward individual national governments, delegations, and diplomatic alliances—for example, the Group of Seventy-Seven at the United Nations (G-77), originally a group of 77 developing nations that now has more than 130 members, and JUSCANZ, a caucus of Japan, the United States, Canada, Australia, and New Zealand. These institutional attempts at influencing nations continued through the period of ratification, and they continue today as interest groups continue to use science to attempt to influence the behavior of governments on treaty-related matters.

Not only did existing institutions involve themselves in the complicated politics and negotiation of the CBD, but several emerged as a direct or indirect result of UNCED, or contemporaneously with the global attention directed at the issue of biological diversity. The World Bank created new financing mechanisms, for example, through its Global Environment Facility (GEF). Several of the information-sharing institutions mentioned above emerged, and new foundations and institutional arrangements for the support of biodiversity research emerged worldwide. The Consultative Group on Biological Diversity, established in 1987, is a forum for grantors that seeks to focus attention on the conservation and restoration of biological resources.

Independent scientific organizations arose in the context of the CBD and continue to date. The National Biodiversity Institute (INBio) of Costa Rica (http://www.inbio .ac.cr) is one example of how difficult it is to categorize institutions. It is a nationally supported (by Costa Rica) not-for-profit private organization with an international advisory board, created in 1989 with the assistance of seed funding from a pharmaceutical company (Merck), in an arrangement catalyzed by a US academic (Cornell University biology professor Thomas Eisner). Its goals span the furthering of science (taxonomy), the development of a new scientific practice area (parataxonomy), the conduct of research, and the application of scientific knowledge on biodiversity to economic activities such as ecotourism, medicine, and agriculture.

Much of the reason for the emergence of new institutions or the transformation of old ones is the inherent complexity of the concept of biological diversity. The definition of biological diversity is so broad that, although it is intrinsically understandable in the abstract, the specifics of its characterization can be maddeningly difficult. Especially as the term *biological diversity* became entwined with the neologism *sustainable development*, organizations sought to adjust themselves and their activities to these new ways of looking at the world.

Worldwide it is historically recent that government, business, and nongovernmental organizations have recognized the need to redefine their environmental roles clearly. In the 1960s, governments began to mediate between an environmentally disaffected public and businesses that had not addressed their indirect or long-term impacts on human welfare. Businesses often reorganized activities to respond to government regulations, lawsuits, and direct citizen actions. Nongovernmental groups emerged to influence the debate, evolved new strategies, and increased their support or use of scientific expertise in new issues in order to level the playing field with both governments and business. Some industry continues to support environmentally related sciences; its activities can sometimes be categorized as "doing well by doing good" (Convis 2001).

The search for underlying models to further develop a science base for environmental management continues. Some have suggested that the Commission for Environmental Cooperation (CEC) of the North American Free Trade Agreement (NAFTA) could serve as a unique model for ensuring that objective science-based environmental concerns remain a major focus in diplomacy (Audley and Vaughan 2003). Debate on similar issues relating to the Central American Free Trade Agreement (CAFTA) continues.

Finally, the issue has been confounded by the very origins of concern over the issues, which resulted in the creation of separate institutions for dealing with each. No one would argue that climate change and biological diversity are not interlinked scientific issues (climate affects the distribution of biota, and biotic distribution is a factor in greenhouse gas cycles) (Peters and Lovejoy 1992). Yet institutional recognition of the need for collaboration and synthesis between the two already defined issues has been slow to develop. The UNCED Secretariat recognized this issue in 2003 by publishing the results of an April 2003 symposium on the relationships between the two issues, held at the University of East Anglia in Norwich, UK (CBD 2003).

Scientific Institutional Implications of Non-Representation

Although the CBD addresses the issue of indigenous people and biological diversity (CBD Article 8(j): Traditional Knowledge, Innovations and Practices), the relationship between these remains problematic, and conflicts between indigenous peoples, scientists, and museums have arisen over collection policies (Chapin and Jutro 1998). Activities have been undertaken to bridge the gaps between indigenous peoples, scientists, and conservation groups. One such effort was undertaken by the Environmental Law Institute with the support of the Office of Research and Development of the US Environmental Protection Agency. An off-the-record workshop on The Relationship among Indigenous Knowledge, Western Science, and Environmental Conservation: Working Together in Collaborative Relationships was held in Front Royal, Virginia, May 18 to 21, 1998. Participants included indigenous representatives, conservationists, and scientists from North America and Latin America (Milius 1998).

Institutional support of field scientific collection has been critical to maintaining and developing science's biotic library. The science of systematics provides the critical information underpinning the understanding necessary to all environmental science; museum collections provide the reference materials that allow scientists to understand ecosystems and their components.

The number of species on Earth is unknown but is immense. As a result, there is no way that representatives of all biota could be either collected or studied. Priorities for collection and research must be established. One of the filters for such collection activity has always been the guidance provided to scientists by indigenous peoples' understanding of their own environments. In the context of both intellectual and real property, this is part of the relationship that has become problematic. This interaction between science, institutions, and indigenous peoples has been addressed in a controversial article critical of large conservation organizations' attention to indigenous issues (Chapin 2004). The relationship is important and is illustrated in a publication of the

National Geographic Society and the Center for the Support of Native Lands. This group of maps illustrates trends in the relationship between indigenous peoples and natural ecosystems in Central America and southern Mexico (National Geographic Society 2002). These maps indicate a strong correlation between indigenous presence and the survival of natural ecosystems; natural ecosystems have a better chance of survival when indigenous peoples inhabit them. The indigenous peoples are, in turn, dependent upon these natural ecosystems. The remaining question is whether indigenous knowledge is useful to environmental research and ecological understanding. Examples of this relationship exist, but evidence of wider usefulness awaits further investigation. Such investigation would have to be multidisciplinary, in part because social science methods would have to be used to gather ecological information (Huntington 2000). Should this hypothesis be supportable, then its corollary would be that maintenance of such knowledge that scientists need is dependent upon the survival of the indigenous peoples.

Given that much institutionally sponsored global change research actually deals with the ecological consequences of change, awareness of this politically fraught issue is of critical importance to both the supporting institutions and the global change scientists. For a concise historical analysis of the indigenous knowledge issue, often referred to as TEK, or traditional ecological knowledge, see Berkes (1993).

Future of Institutions and Science

Today's institutional environment would be unrecognizable to environmentalists of a few decades ago. The field has changed. It has expanded both its spatial and temporal analytic framework for evaluating human influences on the planet. It has emerged from local pollution and natural resource concerns and is maturing to a science based on an appreciation of complexity and the interrelationships between the natural environment, people, and institutions at the regional, transboundary, and global levels. Institutions that wish to support and benefit from science must continue to evolve a better understanding of not only natural environmental interdependencies, but also the evolving relationships between them and the scientific enterprise. The environmental motto used to be "think globally, act locally." A wave of international environmental diplomacy transformed the message to "think locally, act globally." Institutions must now adapt and understand a more complicated environment where the motto has become "think and act locally, think and act globally, and understand the consequences."

Literature Cited

Aristotle. 1952. *Meteorologica.* Cambridge, Mass.: Loeb Classic Library, Harvard University Press.

Aristotle. 1965. *History of animals.* Cambridge, Mass.: Loeb Classic Library, Harvard University Press. (no. 437−9, vol. 9)

Aristotle. 1970. *History of animals.* Cambridge, Mass.: Loeb Classic Library, Harvard University Press. (no. 437–9, vol. 10)

Aristotle. 1991. *History of animals.* Cambridge, Mass.: Loeb Classic Library, Harvard University Press. (no. 437–9, vol. 11)

Audley, J., and S. Vaughan. 2003. *Time for the NAFTA environmental watchdog to get some teeth.* Web commentary, June 24, 2003. Washington, D.C.: Carnegie Endowment for International Peace. (http://www.carnegieendowment.org/publications/index.cfm?fa =view&id=1300&prog=zgp&proj=zted)

Berkes, F. 1993. Traditional ecological knowledge in perspective. In Inglis, J.T., ed. *Traditional ecological knowledge: Concepts and cases.* Ottawa: Canadian Museum of Nature/International Development Research Centre. (pp. 1–9)

Carson, R. 1962. *Silent spring.* Boston: Houghton Mifflin.

CBD Ad Hoc Technical Expert Group on Biological Diversity and Climate Change. 2003. *Interlinkages between biological diversity and climate change: Advice on the integration of biodiversity considerations into the implementation of the United Nations Framework Convention on Climate Change and its Kyoto Protocol.* Montreal: Convention on Biological Diversity. (Technical Series no. 10)

Chapin, M. 2004. A challenge to conservationists. *World Watch Magazine,* November/December: 17–31.

Chapin, M. and P. Jutro. 1998. Native coral. *Science* 281(5384):1805.

Convis, Jr., C. L. 2001. *Conservation geography: Case studies in GIS, computer mapping, and activism.* Redlands, Calif.: ESRI Press.

Huntington, H. P. 2000. Using traditional ecological knowledge in science: Methods and applications. *Ecological Applications* 10(5):1270–4.

Linnaeus, C. 1735. *Systema naturae.* Facsimile Edition. Utrecht, Netherlands: Hes & De Graaf. (2003)

Milius, S. 1998. When worlds collide: Why can't conservation scientists and indigenous peoples just get along? *Science News* 154:6.

National Geographic Society. 2002. *Pueblos Indígenas y Ecosistemas Naturales en Centroamérica y el Sur de México.* Washington, D.C.: National Geographic Society.

Paley, W. 1802. *Natural theology: Or, evidences of the existence and attributes of the deity, collected from the appearances of nature.* Facsimile Edition. New York: Oxford University Press. (2006)

Peters, R. L., and T. L. Lovejoy, eds. 1992. *Global warming and biological diversity.* New Haven: Yale University Press. (386 pp.)

von Humboldt, A. 1815. *Personal narrative of travels to the equinoctial regions of the new continent, during the years 1799–1804.* Translated from the original German into English by Helen Maria Williams. Philadelphia: M. Carey.

WMO/UNEP Scientific Assessment Panel of the Montreal Protocol on Substances that Deplete the Ozone Layer. 2006. *Executive summary, WMO/UNEP scientific assessment of ozone depletion: 2006.* Geneva: World Meteorological Organization and United Nations Environment Programme.

8

Vulnerabilities of Societies under Global Environmental Change (GEC)

Mike Brklacich, May Chazan, and Andrew Dawe

Vulnerability is a term of such broad use as to be almost useless for careful description at the present, except as a rhetorical indicator of greatest concern.
—TIMMERMAN 1981

Social vulnerability is not a new concept. It has been used extensively in many different contexts, including environmental hazards (Mustafa 1998; Lewis 1999; Cutter 2006), famine (Sen 1981; Devereux 1993; Watts and Bohle 1993; Maxwell 1996), and human health (Kalipeni 2000), and more recently the global environmental change (GEC) science community has also engaged extensively with vulnerability concepts and assessments (Luers et al. 2003; Turner et al. 2003; Hesse et al. 2005; Vogel 2006). Despite this popularity, the application of vulnerability concepts within the GEC communities and elsewhere continues to be hampered by multiple definitions (Adger 2006) even though there have been several calls to reconcile conceptual ambiguities and further develop vulnerability assessment methods (Delor and Hubert 2000, Wisner et al. 2004).

In addition there has been limited success in translating vulnerability assessments conducted by the GEC scientists and others into effective policies that enhance the future well-being of individuals and communities that are most threatened by environmental and other stresses (Walker 2006). Multiple reasons underpin the limited application of vulnerability assessments to policy, including the previously mentioned conceptual ambiguities, partially developed vulnerability tools, and minimal collaboration between the GEC science and international development communities. Clearly there is an urgent need for the GEC community to effectively communicate not just its findings but also the policy relevance and social, economic, political, and environmental implications of its work to the international development community (Caldwell 1990).

The overall purpose of this chapter is to provide a context for discussing the application of vulnerability concepts to GEC issues in the Americas. More specifically, the chapter 1) summarizes the evolution of vulnerability concepts, 2) reviews and appraises selected approaches for estimating vulnerability, and 3) suggests recommendations for advancing the inclusion of social vulnerability in GEC science and enhancing its policy relevance.

Evolution of Human Vulnerability Concepts

Just as it is impossible for us to understand a movie by looking at only one still frame it is impossible for us to understand crime or any other process by looking at an individual event or moment.
—MACLEAN 1986

Social vulnerability to GEC has recently emerged as a priority for guiding policy responses to many issues relating to environmental change (Watson et al. 1998; McCarthy et al. 2001; Kovats et al. 2003; UNEP 2003; UN/ISDR 2003). Much of this research has not been well-grounded in the substantial body of work on social vulnerability relating to environmental hazards, famines, and human health. Gilbert F. White's investigations during the 1940s into human activities in the floodplains of major American river systems represents some of the earliest work on human vulnerability to environmental stress, and it set the foundation for the next four decades of natural hazards research (Mitchell 1989). This early work revealed some of the complexities of how social and environmental systems interact to generate vulnerabilities and hinted that the impacts of hazards depend, not only on the magnitude or frequency of the hazard itself (e.g., the flood), but also on the interactions among multiple stressors and causal factors (e.g., the flood, a densely populated floodplain, a poorly developed emergency response system, and a lack of social security) (Jones 1991; Burton et al. 1993). However, vulnerable people and communities were often portrayed as helpless victims, and their capacity to respond to an environmental hazard, even if very limited, was beyond the scope of this work (Emel and Peet 1989).

More recently human vulnerability research has shifted away from approaches that emphasize the "naturalness" of hazards and disasters to place greater emphasis on understanding the prevailing sociopolitical factors that transform environmental stresses into human disasters (Cuny 1983; Hewitt 1983; Wijkman and Timberlake 1984; Blaikie et al. 1994). The roles of uneven political-economic power, entitlements, enfranchisement, command over resources, and institutional stability in producing vulnerabilities have come to light, and thus vulnerabilities are increasingly shown to be generated by social, economic, and political conditions and linked closely to social inequalities (Watts and Bohle 1993; Mustafa 1998). One widely recognized conceptual framework presents vulnerability as having a "double structure" involving the negotiation between the "external sides" (i.e., longer-term structural factors that determine exposure levels to stressors) and the "internal sides" (i.e., the capacity of individuals and communities to cope with stress) of vulnerability (Chambers 1989; Bohle 2001). In climate change debates, this double structure is often portrayed as the likelihood of experiencing the effects of rising sea levels or extreme weather events (and thus the need to reduce carbon emissions in order to mitigate this "external" exposure) and the capacity to adapt to these changes (and thus the need to facilitate "internal" adaptation) (Kelly and Adger 2000). While frameworks guiding climate change policy frequently articulate this as a dichotomy—as *either* mitigation *or* adaptation—social vulnerability perspectives encourage an understanding of the relationships between exposure to stresses and capacity to respond, as well as recognition of the common drivers of both. Indeed, evidence suggests that the same processes that position some people and groups in harm's way—in marginal, low-lying areas and in precarious, resource-based livelihoods—also limit their options for avoiding adverse outcomes (Mustafa 1998; Bohle 2001).

The recasting of hazards and famines research toward an understanding of the underlying causes and multiple dimensions of human vulnerability has also indicated that external stressors most often expose rather than cause vulnerabilities (Walker 1989; Mustafa 1998; Adger 1999). Hence vulnerability should be cast, not as a residual or a result of extreme weather events or other environmental stresses, but rather as a preexisting condition that is often "unveiled" by such stressors (Vogel and O'Brien 2004). It has also become evident that stressors affect various groups and places unevenly as a result of differentiated underlying vulnerabilities (O'Brien and Leichenko 2000). Furthermore, vulnerability research has expanded to encompass a wider range of stressors operating at many spatial and temporal scales (e.g., globalization, GEC, and HIV/AIDS) (Chen and Narasimhan 2003). Also it is now recognized that vulnerabilities are negotiated within contexts of continuous change, where time and space lags between the sources of change and the downstream effects are growing (Kelly and Adger 2000).

The evolution of social vulnerability research in the areas of famines, hazards, and health provides a useful starting point for enhancing the policy relevance of GEC assessments. Overall, contemporary research emphasizes the following with respect to social vulnerability:

- It is defined as the capacity of individuals and communities to respond to (cope with, recover from, and adapt to) external stresses placed upon their livelihoods and well-being.
- It is part of all human systems; it is *exposed* rather than caused by external stressors, with repeated impacts increasing future vulnerabilities.
- It is *differentiated* within and between places and groups; it is linked with broader issues of social, economic, and political inequalities.

- It is generated by the convergence of many drivers and negotiated within *dynamic* societal systems.
- It is reduced by preventing exposure and improving response options to current stresses, in turn potentially reducing both present-day impacts and vulnerabilities to future stressors.

Review and Appraisal of Selected Approaches for Estimating Social Vulnerability

Calculations of vulnerability at the national scale may show comparatively low levels despite there being some highly vulnerable sub-populations or individuals within that nation.
—UNEP 2003

While vulnerability concepts have advanced considerably over the past 15 years, estimating social vulnerability in rapidly changing environments and establishing firmer links between vulnerability science and GEC policy remains problematic. This section provides a review of three approaches for estimating vulnerability and appraises these approaches relative to contemporary vulnerability concepts presented in the previous section. The three selected approaches included in this chapter—the Famine Early Warning System (FEWS), ProVention Consortium: Belize Case Study (PC-B), and Risk Management Assessment Program (RiskMAP)—collectively illustrate approaches with the following characteristics:

- They employ spatial scales ranging from households to nations and temporal scales ranging from a few years to several decades.
- They have been developed and applied to a range of social vulnerability themes (food, risk management, natural hazards).
- They are relevant in the Latin American–Caribbean context.
- They are based on both top-down (FEWS and RiskMAP) and bottom-up (PC-B) approaches.
- They have been designed to address issues relating to public policies and/or programs.

Famine Early Warning System (FEWS)

The US Agency for International Development initiated FEWS and its Information Network (FEWS NET) (www.fews.net) with the overall goal of strengthening the abilities of African countries and regional organizations to manage threats of food insecurity through the provision of timely early warning and vulnerability information (Boudreau and Coutts 2002). FEWS NET is an amalgam that includes several US government agencies, 22 governments in sub-Saharan Africa, UN agencies, nongovern-

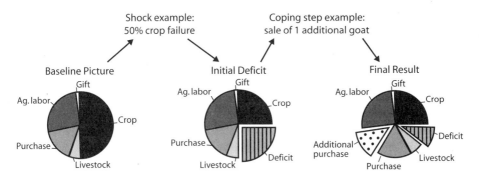

Figure 8.1. FEWS baseline and vulnerability assessments

mental organizations, and the private sector. The FEWS approach has recently been introduced in the Latin America and the Caribbean (LAC) region (Table 8.1), where it has quickly become a central information pillar for government and multilateral decision making on food aid and humanitarian assistance.

FEWS employs a food economy or livelihoods approach, derived from Sen's (1981) work, which expanded food security concepts beyond food supply issues and stressed that access to food was a critical component of food security. The food economy approach (FEA) commences with a baseline assessment that employs household surveys to develop livelihood profiles. These profiles establish how representative households (i.e., with low, medium, or high income) in different food economy zones (FEZs) access food and income and how income is spent. This baseline becomes the context against which the expected effects of a shock or stress to future food security are measured, taking into account the degree to which coping and other strategies can expand to meet basic food requirements and other needs. Figure 8.1 illustrates the application of the FEA to a hypothetical shock (i.e., 50 percent crop failure). FEWS NET routinely tracks monthly trends in food prices, crop and livestock production, weather conditions, trade, and employment. Based on this information, monthly reports detailing food security alerts for each representative livelihood profile in each FEZ of each country are distributed to national and regional decision makers and planners. A four-level food alert system is employed. *Emergency* signifies a significant food security crisis is occurring. *Warning* indicates a food crisis is developing. *Watch* indicates a food security crisis is possible. *No Alert* is used when there are no indications of a food security problem. The baseline livelihood profile provides a snapshot of the current context and thereby captures structural factors that have shaped current levels of food (in)security. The approach illustrated in Figure 8.1 has been used to evaluate food security prospects under a range of scenarios representing plausible futures two years beyond the baseline.

Table 8.1. Comparative assessment of three approaches for assessing societal vulnerability to global change

	Famine Early Warning System (FEWS)	ProVention Consortium: Belize (PC-B)	Risk Management Assessment Program (RiskMAP)
Focus	Food security.	Hazard management/preparedness.	Hazards.
Coverage	Africa: southern Africa, west Africa & Greater Horn of Africa. Mesoamerica (MFEWS): 4 countries commencing 2004–05.	Municipal (Belize City case study).	Latin America & Caribbean.
Lead Agency	USAID.	ProVention Consortium/ Red Cross.	Inter-American Development Bank (IDB).
Application	Early warning & emergency food aid planning.	Disaster preparedness, prevention, and mitigation.	Disaster prevention and response; development planning.
Spatial Units	Households in food economy zones.	Municipalities to nations/states.	Nations/states to municipalities.
Temporal Units	5 years (+/- 2 years from present).	>5 years.	Several/multiple decades.
Methods	Baseline livelihood profiles derived from household surveys. Shocks from trade, agricultural production, food price. Coping based on livelihood options.	Community-based participatory workshops that document threats to human well-being and identify vulnerabilities, capacities, and mechanisms to improve disaster preparedness.	4 complimentary indices, with 3–4 factors, each with up to 8 indicators. Hazard frequency and magnitudes; disaster impacts; exposure, fragility and resilience; response capacity.

Data	Baseline livelihood profiles (food & income sources, assets held). Routine production, price & trade data.	Baseline vulnerabilities and institutional response capacity. Actions to reduce risk, improve rehabilitation and hazard responses.	Routine long-term data. DESINVENTAR disaster database. Human development indices.
Scenario	External shocks either hypothetical or from field reports.	Current natural and anthropogenic hazards, institutional capacities.	Hazard return frequency, empirical & hypothetical, response capacity.
Application to GEC–Vulnerability	Derived from food security concepts. Livelihoods capture current constraints. Timeline shorter than GEC. Need representative households. Plausible coping strategies.	Derived from contemporary hazard, vulnerability and risk concepts. Considers threats separately, does not consider cumulative threats. Timeline shorter than GEC. Bridges local response capacity and environmental processes, connections to broader scales less developed.	Derived from disaster, hazard and development concepts. Macro and institutional measures highlight fiscal constraints. Variable timelines, trends and return rates. Uses available data. Comprehensive structural coverage.
Further Information	Boudreau & Coutts 2002 www.fews.net	Provention Consortium 2005 www.proventionconsortium.org	Cardona 2005 www.iadb.org

FEWS and Advancing the Relevance of Social Vulnerability
in GEC Research

FEWS offers several opportunities for advancing the relevance of social vulnerability science within a GEC policy context. FEWS recognizes entitlements and access as crucial elements of food security. This represents a significant advance over much of the earlier GEC-vulnerability research, which has often focused on food supply as the chief indicator of food security. It also reflects the conceptual advances summarized at the start of this chapter by focusing attention on underlying social, political, and economic causes of vulnerability. In addition, with baseline livelihood profiles anchored in household surveys, FEWS has the added advantage of capturing the extent to which current institutions shape food security. Furthermore, the baseline livelihood profiles consider how coping strategies could be employed to counter shocks, further mirroring the social vulnerability concepts outlined above.

Some recent work has indicated FEWS has played an important role in identifying food gaps and guiding food aid, but FEWS has been less successful in penetrating the multiplicity of underlying stresses (e.g., economic shocks, HIV/AIDS, and growing poverty) that drive food insecurity in the southern African region (Drimie and Vogel 2005). In addition, FEWS employs a timeline of approximately five years, which is considerably shorter than the time frames associated with GEC. However, as one of the most systematic and comprehensive information tools available to decision makers, it has become widely integrated in Central America, and FEWS is positioned to play an important role in the development of longer-term policy responses that simultaneously consider social vulnerability and GEC stresses.

ProVention Consortium: Belize Case Study (PC-B)

The ProVention consortium (www.proventionconsortium.org) includes governments, international organizations, academic institutions, the private sector, and civil society organizations (CSOs) working toward an overall goal of supporting developing countries in their efforts to reduce the risk and social, economic, and environmental impacts of natural and technological disasters on the poor (ProVention Consortium 2005). Activities are aimed at advancing disaster management policies and programs and include identifying risks to human well-being, reducing vulnerabilities by enhancing response capacities, and protecting communities and investments from disasters (Table 8.1).

The Belize vulnerability and capacity assessment case study provides insight into the ProVention Consortium's bottom-up approach to enhance and secure human well-being from a variety of threats. PC-B engages community leaders drawn from national and local governments, CSOs, the private sector, and academic institutions in participatory vulnerability and capacity assessment (VCA) processes. Table 8.2 presents the basic steps in the PC-B VCA. It commences with an articulation of the threats, including environ-

Table 8.2. ProVention consortium vulnerability and capacity assessments: Examples derived from the Belize case study

Major Phase	Floods	HIV/AIDS
1. Clarify risks, vulnerabilities, & capacities.	Infrastructure & livelihoods are at risk. Squatters in low-lying coastal communities are most vulnerable. Capacity is constrained by limited training, medical support, & disaster planning.	Livelihoods, lives, & work force are at risk. Ignorance, discrimination, & peer pressure heighten vulnerabilities. Capacity is constrained by community acceptance of HIV/AIDS & lack of awareness.
2. Transform vulnerabilities into capacities.	Upgrade infrastructure, agricultural practices, & community life-saving skills. Develop & disseminate disaster plan. Relocate communities from low-lying areas.	Improve awareness. Enhance health care access for testing & treatment. Improve services for sex trade workers.
3. Identify prevention & preparedness options.	Enforce construction code & land use zoning regulations. Identify safe & dangerous areas. Improve knowledge of response plans.	Develop awareness programs for teens, parents, teachers, etc. Increase support & coordination across health & counseling agencies. Improve social services for families with HIV/AIDS members.
4. Take concrete implementation actions.	Clean & maintain drains. Install early warning systems. Community participates in disaster response exercises. Housing inspected to ensure codes enforced.	Campaign to improve awareness. Secure resources & personnel for testing, health care, and delivery of social services to affected families.

Derived from ProVention Consortium (2005) and www.proventionconsortium.org

The Belize case study included assessments of other threats including fire and hurricanes. Floods and HIV have been selected for inclusion here to illustrate threats that are directly linked to natural hazards as well as a threat that is not directly linked to the environment and GEC.

mental and non-environmental stresses (e.g., floods, hurricanes, HIV/AIDS), what is threatened (e.g., human livelihoods, infrastructure, human lives), factors contributing to vulnerabilities (e.g., poverty, discrimination, ignorance), and current capacities to reduce vulnerabilities. Once this initial context is set, the process switches to investigating opportunities to reduce negative impacts stemming from each of the threats, including the identification of disaster preparedness options and the development of priorities for policies and programs to address vulnerabilities. Overall PC-B strives to move beyond identifying the vulnerability-hazard couplings that underpin human disasters to identifying options for building new community level capacity to recognize, cope with, and adapt to threats and thereby enhance human well-being.

PC-B AND ADVANCING THE RELEVANCE OF SOCIAL VULNERABILITY IN GEC RESEARCH

PC-B puts into operation many of the fundamental concepts that characterize contemporary social vulnerability research. It recognizes that vulnerabilities are generated by underlying sociopolitical factors and exposed by stresses such as floods and hurricanes and that vulnerabilities vary over space and time. In addition, PC-B emphasizes the need to consider a wide range of environmental and non-environmental stresses that can impinge upon human well-being and moves beyond identifying vulnerabilities to building response capacity. Like FEWS, however, the application of PC-B beyond a five-year timeline remains unknown and needs to be considered before it can facilitate GEC-vulnerability science. In addition, PC-B is designed to consider multiple threats to human well-being, but these threats are treated sequentially, and the cumulative effects of multiple threats appear to be beyond the PC-B approach. From a policy and program perspective, the PC-B engagement of a wide range of stakeholders in all phases of the VCA, ranging from the initial articulation of vulnerabilities to the development of disaster reduction plans, increases the likelihood of successful implementation of disaster reduction strategies. This successful stakeholder engagement in policy and programming also provides an opportunity for more locally informed and sensitive GEC-vulnerability science that incorporates nonexpert participation.

Risk Management Assessment Program (RiskMAP)

RiskMAP (www.iadb.org) is a complementary series of four indices developed by the Inter-American Development Bank for the identification and management of disaster risk (Cardona 2005). In this framework, vulnerability is the focus for action to reduce disaster risk and bolster disaster prevention (Table 8.1). This indicator-based approach to vulnerability highlights awareness of the connections between disaster risk reduction and development policies and planning. (See Figure 8.2.)

Four indices are employed and collectively provide a comprehensive approach to identifying risk factors for both large- and small-scale impacts (in terms of frequency,

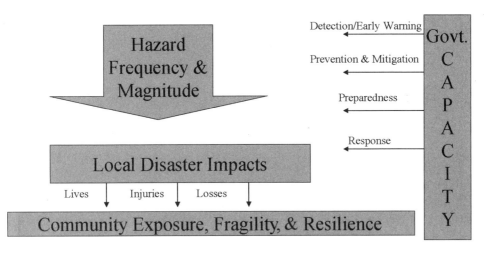

Figure 8.2. Overview of Risk Management Assessment Program (RiskMAP)

magnitude, exposure, coping, and resilience) and weaknesses in response and recon-struction capacities. The disaster deficit index focuses on a country's macroeconomic and financial ability to cope with a situation, as well as the historical trends and likeli-hoods of impacts of different scales upon the national economy. It uses deductive indi-cators and simple physical risk modeling to estimate the likelihood (risk) of extreme haz-ard events. The local disaster index "identifies the social and environmental risks resulting from more recurrent lower level events (which are often chronic at the local and subnational level)" (Cardona 2005). It examines past impact events at differing scales and levels. The prevalent vulnerability index "depicts vulnerability conditions by measuring exposure in prone areas, socioeconomic fragility and lack of social resilience" (Cardona 2005). The risk management index examines the specific measures taken within a country to prepare for, mitigate, and respond to disaster impacts, including organizational and institutional capacity and planning for action, using a composite of aggregate quantitative and qualitative indicators.

RiskMAP and Advancing the Relevance of Social Vulnerability
in GEC Research

The three factors comprising the RiskMAP vulnerability index—exposure, fragility, and resilience—are consistent with political economy/ecology approaches to understanding hazards and disasters and with the double structure of vulnerability discussed earlier in this chapter (Bohle 2001; Wisner et al. 2004). By comparison with the two frameworks described above, the consideration of hazards in the RiskMAP approach, with its long time frames, makes it highly relevant within a GEC context. It incorporates climate change impact and adaptation considerations through its attention to both hazards and vulnerabilities. The policy relevance for GEC is increased by its further consideration of (fiscal) capacity in responding to global change, while identifying the social and geographic areas of vulnerability that are to be reduced through adaptation activities and long-term development planning. This focus on operative capacity, however, comes at the cost of lost attention to individual and group agency beyond the purely economic realm. Finally, the extent of data and indicators used in these indices should not be assumed to produce precise results—the level of aggregation masks as much about the distribution of vulnerability as is revealed by the correlations of factors. This concern with data is present throughout GEC science, and macro-level policy making would be well complemented by the kind of bottom-up input described in the PC-B case.

Advancing the Relevance of Social Vulnerability to GEC

The significant advances in vulnerability science in the hazards, famine, and health fields over the past 15 years provide a stable foundation for the next generation of GEC-vulnerability research. It is crucial that the GEC communities continue to build vulnerability tools from the ground up, with analytical frameworks derived from innovative conceptual development and empirical studies, rather than driven solely by data availability. This would ensure that future research focuses on the multiple and often entangled processes that shape social vulnerability. Vulnerability needs to be understood as part of everyday life and not as an anomaly caused by an environmental stress.

In addition, while it is useful to conceptualize social vulnerability according to its external and internal dimensions (i.e., the structural conditions that can increase exposure to stresses and the capacities of individuals and groups to cope and respond, respectively), there must be research that moves beyond such dualisms to show how these "sides" interact in particular locations. Finally, most current approaches for estimating social vulnerability operate at one scale and often employ formulaic or highly generalized frameworks. Flexible frameworks that can capture how global processes enter, operate, and are negotiated within communities, as well as frameworks that delineate and weight factors driving community level social vulnerability, will be key to advancing GEC-vulnerability science.

GEC-Vulnerability Policy Perspectives

Enhancing the policy relevance of GEC-vulnerability science requires a fundamental shift from current practices, which regularly view biophysical and social scientists as external to policy stakeholders, to practices that view the scientific and policy communities as collaborators. This approach will facilitate the development of common and shared goals and thereby assist with a smoother transition from experimental or prototype tools to analytical frameworks that can be readily incorporated into institutional settings. This process needs to commence with an institutional mapping exercise, to identify the agencies and actors responsible for GEC and related policies, and an assessment of the relationships amongst these agencies and actors. The vulnerability-GEC science communities need to understand the prevailing policy context and address several questions: How are social vulnerability questions framed? What sorts of new information or evidence are needed, and when? It is critical that strengths and weaknesses of response options, including the "do nothing" option, be assessed? Once these questions have been answered and the context has been established, the social vulnerability-GEC community will have a stronger basis for interacting with its policy partners on a continual basis. This will facilitate the development of a reflexive approach that will be able to monitor and respond to shifts in both the science and policy communities.

Acknowledgments

This paper benefited from funding provided by Canada's Social Sciences and Humanities Research Council (www.sshrc.ca), the International Council for Science (www.icsu.org), and the International Human Dimensions Programme on Global Environmental Change (www.ihdp.org) and is derived from research conducted in conjunction with the Global Environmental Change and Human Security (www.gechs.org) and the Global Environmental Change and Food Systems (www.gecafs.org).

Literature Cited

Adger, N. 1999. Social vulnerability to climate change and extremes in coastal Vietnam. *World Development* 27:249–69.

Adger, N. 2006. Vulnerability. *Global Environmental Change* 16:268–81.

Blaikie, P., T. Cannon, I. Davis, and B. Wisner. 1994. *At risk: Natural hazards, people's vulnerability and disasters.* London: Routledge.

Bohle, H.-G. 2001. Vulnerability and criticality: Perspectives from social geography. *IHDP Update* 2/2001. Bonn: IHDP.

Boudreau T., and P. Coutts. 2002. *Food economy in situations of chronic political instability.* Working Paper 188. London: Overseas Development Institute.

Burton, I., R. Kates, and G. White. 1993. *The environment as hazard,* 2nd ed. New York: Guilford.

Caldwell, L. K. 1990. *Between two worlds: Science, the environment movement and policy choice.* New York: Cambridge University Press.

Cardona, O. 2005. *Indicators of disaster risk and risk management.* Summary report. Washington, D.C.: Inter-American Development Bank, Sustainable Development Department.

Chambers, R. 1989. Vulnerability, coping and policy. *IDS Bulletin* 20(2):1–7.

Chen, L., and V. Narasimhan. 2003. Global health and human security. In Chen, L., S. Fukuda-Parr, and E. Seidensticker, eds. *Human security in a global world.* Cambridge, Mass.: Harvard University Press. (pp 183–93)

Cuny, F. 1983. *Disasters and development.* New York: Oxford University Press.

Cutter, S. L. 2006. *Hazards, vulnerability and environmental justice.* London: Earthscan.

Delor, F., and M. Hubert. 2000. Revisiting the concept of vulnerability. *Social Science and Medicine* 50:1557–70.

Devereux, S. 1993. *Theories of famine.* London: Harvester Wheatsheaf.

Drimie, S., and C. Vogel. 2005. *Unravelling the entangled crises: Challenges around vulnerability analysis in the southern African region.* Paper presented at the Sixth Open Meeting of the Human Dimensions of Global Environmental Change Research Community, Bonn, Germany, October 2005.

Emel, J., and R. Peet. 1989. Resource management and natural hazards. In Peet, R., and N. Thrift, eds. *New models in geography,* vol.1. Boston: Hyman. (pp 49–76)

Hesse, A., B. Goebel, L. Mullin, and Y. Tian. 2005. Conference book for the Sixth Open Meeting of the Human Dimensions of Global Environmental Change Research Community, Bonn, Germany, October 2005.

Hewitt, K., ed. 1983. *Interpretations of calamity.* Boston: Allen and Unwin.

Jones, D. 1991. Environmental hazards. In Bennett, R., and R. Estall, eds. Global change and challenge: Geography for the 1990s. New York: Routledge. (pp 27–56)

Kalipeni, E. 2000. Health and disease in southern Africa: A comparative and vulnerability assessment. *Social Science and Medicine* 50:965–83.

Kelly, P., and W. N. Adger. 2000. Theory and practice in assessing vulnerability to climate change and facilitating adaptation. *Climatic Change* 47:325–52.

Kovats, S., K. Ebi, and B. Menne. 2003. *Methods for assessing human vulnerability and public health adaptation to climate change.* In Health and Global Environmental Change Series No. 1. Geneva: World Health Oganization.

Lewis, J. 1999. *Development in disaster-prone places: Studies of vulnerability.* London: Intermediate Technology Development Group.

Luers, A., D. B. Lobell, L. S. Sklar, C. L. Addams, and P. A. Matson. 2003. A method for quantifying vulnerability, applied to the agricultural system of the Yaqui Valley, Mexico. *Global Environmental Change* 13:255–67.

Maclean, B. 1986. Critical criminology and some limitations of traditional inquiry. In Maclean, B., ed. *The political economy of crime: Reading for a critical criminology.* Toronto: Prentice Hall. (pp. 2–12)

Martin, R. 2001. Geography and public policy: The case of the missing agenda. *Progress in Human Geography* 25(2):189–210.

Maxwell, D. G. 1996. Measuring food insecurity: The frequency and severity of "coping strategies." *Food Policy* 21(3):291–303.

McCarthy, J., O. Canziani, N. Leary, D. Dokken, and K. White, eds. 2001. *Climate change: Impacts, adaptation, and vulnerability.* Third assessment report, IPCC Working Group II. Cambridge: Cambridge University Press.

Mitchell, B. 1989. *Geography and resource analysis,* 2nd ed. New York: John Wiley and Sons. (pp. 174–98)

Mustafa, D. 1998. Structural causes of vulnerability to flood hazard in Pakistan. *Economic Geography* 74:289–305.

O'Brien, K., and R. Leichenko. 2000. Double exposure: Assessing the impacts of climatic change within the context of economic globalization. *Global Environmental Change* 10:221–32.

Parson, E. A. 1997. Informing global environmental policy-making: A plea for new methods of assessment and synthesis. *Environmental Modeling and Assessment* 2: 267–79.

ProVention Consortium. 2005. *Community risk assessment and action planning project: Belize Red Cross vulnerability and capacity assessment workshop.* http://www.provention consortium.org/themes/default/pdfs/CRA/Belize.pdf

Sen, A. 1981. *Poverty and famines: An essay on entitlement and deprivation.* Oxford: Clarendon Press/Oxford University Press.

Timmerman, P. 1981. *Vulnerability, resilience and the collapse of society.* Environmental Monograph No. 1, Institute of Environmental Studies, University of Toronto.

Turner, B. L., R. E. Kasperson, P. A. Matson, J. J. McCarthy, R. W. Corell, L. Christensen, N. Eckley, J. X. Kasperson, A. Luers, M. L. Martello, C. Polsky, A. Pulsipher, and A. Schiller. 2003. A framework for vulnerability analysis in sustainability science. *Proceedings of the National Academy of Sciences of the United States of America* 100:8074–9.

UNEP. 2003. *Assessing human vulnerability to environmental change: Concepts, issues, methods and case studies.* Nairobi: Division of Early Warning and Assessment, UNEP.

UN/ISDR. 2003. Living with risk: Turning the tide on disasters towards sustainable development. Geneva: United Nations Inter-Agency Secretariat of the International Strategy for Disaster Reduction.

Vogel, C. 2006. Foreword: Resilience, vulnerability and adaptation: A cross-cutting theme of the International Human Dimensions Programme on Global Environmental Change. *Global Environmental Change* 16: 235–6.

Vogel, C. and K. O'Brien. 2004. Vulnerability and global environmental change: Rhetoric and reality. *AVISO* No.13. Global Environmental Change and Human Security Project. Carleton University, Ottawa.

Walker, P. 1989. *Famine early warning systems: Victims and destitution.* London: Earthscan.

Walker, P. A. 2006. Political ecology: Where is the policy? *Progress in Human Geography* 39:382–95.

Watson, R., M. Zinyowera, R. Moss, and D. Dokken, eds. 1998. *The regional impacts of climate change: An assessment of vulnerability.* A Special Report of the Intergovernmental Panel on Climate Change Working Group II. Cambridge: Cambridge University Press.

Watts, M., and H.-G. Bohle. 1993. The space of vulnerability: The causal structure of hunger and famine. *Progress in Human Geography* 17:43–67.

Wijkman, A. and L. Timberlake. 1984. *Natural disasters: Acts of God or acts of man?* London: Earthscan.

Wisner, B., P. Blaikie, T. Cannon, and I. Davis. 2004. *At risk: Natural hazards, people's vulnerability and disasters,* 2nd ed. London: Routledge.

9

What Social and Natural Sciences Could Learn from Each Other: The Challenge of Interdisciplinarity

Barbara Göbel

Global Environmental Change as an Exemplary Case for Interdisciplinary Science

Global environmental change (GEC) research provides the opportunity to assess the potentials, limitations, and challenges of interdisciplinary approaches between natural and social sciences (including humanities). GEC can be defined as the set of biophysical transformations of land, oceans, and atmosphere driven by an interwoven system of human and natural processes. Human activities are having a measurable effects on Earth's environment. Nearly 50 percent of the land surface has been transformed by direct human action, with significant consequences for biodiversity, soils, and climate. More than half of the accessible freshwater is used directly or indirectly by societies, and underground water resources are being depleted in many areas. The atmospheric concentrations of several important greenhouse gases have increased substantially since industrialization, with potentially dramatic consequences for Earth's climate. Coastal and marine habitats are being altered at a rapid rate, while fisheries worldwide are being depleted. Therefore it is becoming imperative to better understand the dynamics of land-use and land-cover changes, the impacts of environmental changes on coastal zones, and the processes of urbanization, as well as the transformations of water systems and of the global carbon cycle. In order to do so, we have to recognize that global environmental changes are intimately connected with processes of socioeconomic, political, and cultural change at scales ranging from the local through to the global. This chapter presents a global perspective and focuses on intersections between GEC and the human dimensions of globalization, that is, it considers globalization research with a focus on the environmental dimensions of globalization.

Globalization is the growing and accelerated interconnectedness of the world, which has been fueled by enhanced flows of goods, capital, information, and people, resulting in the world's becoming more and more a single place—a "global village"—in which distant peoples share knowledge and lifestyles, and different institutions function as parts of one complex system. Although globalization processes (as well as global environmental changes) are not new phenomena in human history, the scale, quality, and speed of the globalization of the last decades are without precedent. People increasingly perceive local situations and events as being influenced by external, often unpredictable, factors that they can control only in very limited ways. At the same time, there is a growing recognition that local events or actions in one place may have consequences in other locales worldwide, with often unforeseeable results and diffuse feedbacks. The everyday experience of an increasing degree of interconnectedness between local, national, regional, and global dimensions, which cannot be grasped by the individual in its whole complexity and dynamic, is restructuring basic relations between humans and between humans and their environment. The complexity, connectedness, and uncertainty of globalization make it very difficult to keep track of the intricate webs, effects, feedbacks, and anticipations that underlie the dynamics of globalization and global environmental change.

GEC and globalization have growing impacts on human decision making and human interactions at the local and regional levels, and both are in turn influenced by local and regional processes. Therefore human security, social and cultural well-being, and economic development cannot be addressed without taking GEC and globalization into account and without understanding the complex interlinks and dynamics between global, regional, and local scales.

GEC is imposing new demands upon political decision makers as well as the scientific community, and the setting of local decision-making and scientific communities in a global context is posing new challenges. One main challenge is how to respond to the diverse, often ambivalent and contested requests societies pose to science: the provision of codified, validated, and certified data; explanations of complex phenomena; future predictions and scenarios; and applicable solutions. Related to this is the challenge to produce scientific knowledge and to develop modes of producing, synthesizing and translating scientific knowledge that consider political and societal processes.

It has been stressed recently that GEC researchers who want their findings to become more relevant to society should take three interconnected dimensions into account: 1) interdisciplinarity, 2) a problem-solving orientation that considers the applicability of the scientific insights, and 3) stakeholder involvement in the design and the translation of scientific knowledge into other arenas of action. The combination of these three dimensions is seen as a prerequisite for developing the kinds of mitigation or adaptation strategies to cope with the impacts of GEC that are feasible from social, cultural, political, and economic points of view.

GEC is characterized by a complexity of coupled socio-environmental processes whose multiple temporal and spatial dimensions are connected in such a way that they influence each other. Therefore it is necessary that different scientific communities develop common questions, integrate their concepts, and design converging research strategies. Disciplinary perspectives that allow only partial and fragmented understanding of GEC must be augmented by more holistic and integrative analytical frameworks. Socioeconomic systems can never be considered independent of their environments and therefore demand the integration of social science approaches (putting *people* at the center of the analysis) with natural science approaches (putting *nature* at the center of the analysis).

The increasing urgency of the matter, as more and more negative impacts of GEC are perceivable in many parts of the world, demands that fundamental research is complemented by research oriented toward problem solving. This can be achieved only if a stronger nexus is built between science and society. The production of scientific knowledge has to be embedded in broader societal processes. In order to link science and practice, research has to be organized in ways that consider societal demands, link the scientific knowledge to other types of knowledge, and translate it into other arenas of action. This is not an easy or uncontested process since it requires new ways for the negotiation of reciprocal knowledge transfer.

Challenges of Interdisciplinarity

Disciplines are historically grown units. They are developed in order to converge elements of understanding and explanation by unifying perspectives, by constructing metalanguages that link the search for regularities and patterns and limit the research to questions within the range accepted by the scholars involved, and by discarding apparent noise. As Sander van der Leeuw, of Arizona State University, in Tempe, states, "Disciplines are a question of . . . discipline."

Differences exist between and within disciplines with regard to many areas:

- data (e.g., quantitative or qualitative data)
- methodology (e.g., the roles of induction, deduction, models, narratives or case-study–based descriptions, laboratory or field observation)
- levels of generality (e.g., fundamental, applied, or practical research)
- modes of generalization (e.g., objective or context-dependent or relative)
- degrees of precision (e.g., quantification and isolation of information or conceptual fuzziness and focus on the connectivity between information bits)
- scales at which each discipline highlights different phenomena (e.g., global, regional, or local)
- cultural styles (e.g., epistemological traditions, the respective roles of theory and observation, the role of scientific knowledge with regard to expert knowledge or indigenous knowledge)

- organizational settings of the scientific practice (e.g., academic structures and their settings in the institutional landscape for knowledge production, socialization of academic and career systems, institutional organization of knowledge as the cause of disciplinary fragmentation or transdisciplinary settings)
- degrees of investment (e.g., scattered and diverse investments or large investments in mega-projects, many researchers on small topics or few researchers on large ones)

These differences lead to competition and conflicts between disciplines; they also set the dividing line between natural and social sciences.

What does this mean for the research on GEC? First of all, it is important to make the contexts of knowledge production more explicit. Only the recognition of these contexts and their historical development will allow us to overcome their limitations and address a problem with the great complexity of GEC. Only such a recognition will allow us to transcend disciplinary boundaries without losing disciplinary roots. It will broaden scientific agendas historically negotiated between only certain groups of scientists (e.g., from the "North") and open a new terrain of knowledge production that can take advantage of the heuristic and institutional experiences accumulated through time. We should recognize that scientific communities dealing with environmental matters have different disciplinary roots, constituencies, interests, and theoretical and methodological approaches. They also have been socialized with different institutional backgrounds and scientific practices. They focus on different aspects or dimensions of the environment, address different environmental problems, and frame questions in different ways. Consequently they search for different data and provide different types of answers.

Within science a great diversity of perspectives on human-environment relations has developed. The different perspectives are based on different conceptualizations of basic components of reality (e.g., different perspectives on space and time as well as different basic ontologies such as subject-object relations). Depending on their disciplinary and paradigmatic backgrounds, scientists use different concepts to address the "environment." For example, the differences in concepts between the German *Natur, Umwelt,* and *Mitwelt* (nature, environment, and surroundings) or between the French *environnement* and *milieu* indicate fundamental differences with regard to the basic relation between humans and nature.

Natural sciences often use an *ecosystem* approach for their integrative analysis of human-environment relations. In this approach, human beings (without greater differentiation between individuals or groups) are looked upon as a component of an ecosystem, linked via feedback loops to other (natural) components of the system. The strength of the ecosystem approach is its potential to systematize, organize, and reduce complexity by focusing on key components and on basic relations between these components. This generates testable and quantifiable hypotheses on some aspects of the dynamics of coupled systems. The box "human dimensions" remains, however, often quite vague and mechanistic in a sort of "behavioristic action-reaction" manner. An analysis of the development of the so-called human dimensions in the GEC programs shows that the box "human dimensions" was filled at first with the factor "population," then "institutions" were added, and later several factors were often combined without their theoretical roots taken into account. Eclectic combinations such as "demography," "institutions," "economy," and "knowledge/religion/values" are quite common. In the ecosystem approach, the box "human dimensions" remains fragmented. It depicts a fragmented, static, and partial picture of societies, in which many important social, economic, political, and cultural dimensions are not included. In this ecosystem approach, "social life" can hardly be grasped in its whole complexity, dynamic, and fuzziness.

When social sciences address human-environmental relations, they put *people* in the center of the analysis. They focus on agency and the economic, social, political, and cultural contexts for decision-making processes (e.g., the relation between power, knowledge, and action), looking at the interrelations between action and structure as well as the interactions between individuals and groups. One of the common shortcomings of social science approaches to human-environment relations is that the natural environment is taken as given or is reduced to the provision of resources or ecosystem goods and services. Therefore one of the challenges for social science is to overcome the static perspective on the natural environment and take into account environmental changes, that is, the broader dynamics between the geosphere, biosphere, and atmosphere.

In its current stage of development, global change research is somehow located between these two positions. This is related to the history of research on global environmental change. The GEC research agenda has been driven mainly by natural science, which has begun to include more and more social science perspectives as the understanding of the problem has advanced and political pressures have mounted. So far, the input of social sciences is limited to a certain range of concepts that lend themselves to cooperation with natural sciences. Examples are rational choice approaches as game theory, new institutional economics, and decision-making theory or systemic approaches linked to structural-functionalism or systems theory. A greater systematic effort is needed to identify ways of including other theoretical approaches, such as the analysis of the whole cultural dimension or of power relations, which could make an important contribution to the GCR agenda.

The process of developing interdisciplinary research agendas is one of negotiation. These negotiations are embedded in power structures and are influenced by scientific and cultural styles. Mutual trust has to be built through developing win-win situations. In addition to investments in trust-building mechanisms, other relevant factors for a successful process are the clarification of mutual expectations and goals, the joint development of fair and pragmatic rules of negotiation, and the recognition that diversity, in spite of its implicit tensions, can build a creative space for knowledge production. It seems appropriate to take an ethnographic approach—to try to understand the "other" by stepping out of the disciplinary boxes into which one has been socialized and exposing oneself intellectually to other world views. In the same way as a prerequisite for a fruitful intercultural dialogue is recognition of differences and respect for them, interdisciplinarity requires that all disciplines acknowledge that disciplinary practices may exist in many different but equivalent ways, each with its own requirements. The difficulty is in *not* going for the simple solution. Assessments of interdisciplinary research projects have shown that, for the negotiation of common interdisciplinary agendas, the *process* is as important as the results or the outcomes of the process.

A critical challenge is to negotiate the essential questions to be asked in such a way that they can be answered from many different cultural, disciplinary, and practical backgrounds. The main question is, What can one do to make the various images as compatible and as complementary as possible? That seems a much more rewarding task than attempting to negotiate the concepts to be used in answering those questions— which usually leads to disciplinary disputes and suppresses the fuzziness of concepts that is essential for creativity in an interdisciplinary context.

Future Challenges for Research on Global Environmental Change

In spite of all its important contributions, research on GEC needs to improve in three main areas. It has not been successful in organizing a strong input of social sciences, including humanities, in epistemologically coherent ways. Often incommensurabilities and inconsistencies are created because theoretical approaches are simply added on, basic differences in scientific procedures are not taken into account, and methods are combined without consideration of the different theoretical contexts and paradigms in which they developed. Research on GEC also continues to struggle to find a strong nexus between science and policy (or, in broader terms, practice) in order to make scientific knowledge more relevant to society. Further advances will require the production of knowledge that leads to actions that address social issues and iniquities. This will require GEC research to learn from other scientific communities, such as development research, which has a long history of action-oriented research. Researchers will have to engage multiple cultural perspectives (ontologies and knowledge systems) on GEC (definition, context, causes, impacts, etc.) into one explanatory framework.

To overcome these gaps, three interrelated steps are necessary:

- GEC researchers must reflect on the process of scientific knowledge production (e.g., the relation between knowledge production and power), the negotiation of knowledge, its mediation, and the translation of knowledge into action. At this stage of development of global change research, those involved need to take stock in a more systematic way of the epistemological, theoretical, and methodological achievements of GEC research and its interdisciplinary and international dimensions, in order to identify bottlenecks and to develop strategies to overcome them.
- Research on GEC must be connected in broader and more encompassing ways to the-

oretical and methodological developments in social sciences. Social sciences have made important theoretical and methodological contributions to the study of processes of economic, social, political, and cultural globalization and in the realms of development research, which are of relevance to our understanding of GEC, but so far they have not been taken into account. This would also link environmental change research in more systematic ways to questions of development (e.g., the UN Millennium Development Goals).

- GEC research must be (re)framed in ways that are more useful for political decision makers and stakeholders in general. The final goal should be to build bidirectional processes between the construction of scientific knowledge (including the design and coordination of research) and the dynamics of practice. On the one hand, it is necessary to analyze how global change research should be designed in order to encompass different cultural perspectives and address societal needs (knowledge for action and actionable knowledge), as well as how it should be translated from one arena of knowledge production to another (e.g. knowledge broker, mediation platforms). On the other hand, it is also important to better understand policy processes and to identify the docking points where research could fit, in order to make a more substantial contribution to policy processes.

10

Legal Frameworks and Biodiversity: The Impact of Ownership and Control of Biodiversity on Science

Susanne Reyes-Knoche

Since the Convention on Biological Diversity (CBD) entered into force, biodiversity-related research activities with genetic resources have to go through the access procedure of the CBD and the respective national legislations of the Parties to the Convention. This paper examines the relationship between scientific activities, including global environmental change (GEC) science-driven projects and CBD access provisions. It aims at highlighting the problems arising out of this relationship and provides suggestions that might ease scientific work in this field. The focus is on the access and benefit-sharing (ABS) system of the CBD and its impacts on scientific activities in Latin America.

Endangered Biological Diversity, Natural Resources Sovereignty, and the Convention on Biological Diversity

Biological diversity, or *biodiversity*, means the variability among living organisms from all sources, including terrestrial, marine, and other aquatic ecosystems, and the ecological complexes of which they are part (Article 2 of the CBD).

Biodiversity's importance lies in the fact that it makes up the structure of the ecosystems and habitats that support essential living resources, including wildlife, fisheries, and forests. It plays a substantial role in providing for basic human needs such as food, shelter, and medicine. (See Gehl Sampath 2005, p.13 ff, concerning the importance of biodiversity for drug R&D.) Biodiversity also has recreational, cultural, and aesthetic values.

Over the past years, a significant loss of biological diversity has occurred mainly because of society's growing consumption of natural resources and increasing population. In awareness of the danger to biodiversity and recognizing the vital task of anticipating, preventing, and attacking the causes, 187 countries and the European Community signed the Convention on Biological Diversity in 1992. The CBD entered into force in 1993. The main objectives of this legally binding convention, as defined in its Article 1, are 1) the conservation of biological diversity, 2) the sustainable use of its components, and 3) the fair and equitable sharing of the benefits arising out of the utilization of genetic resources.

Before the CBD entered into force, genetic resources were commonly considered the "common heritage of mankind." Accordingly international law tolerated an open access regime to genetic resources (Stoianoff 2004). The use of genetic resources for the development of new products was largely undertaken without regard to the state, the communities, or other potential stakeholders from which the source material was drawn. Due to the recognition of the increasing economic value of genetic resources, the open access regime was put into question.

With the signing of the CBD, the "common heritage of mankind" doctrine came to an end. Contrary to that doctrine, the CBD recognizes and reiterates the applicability of a generally accepted sovereign right of states to exploit their own resources pursuant to their own environmental policies explicitly for genetic resources (Stoll 2004; CBD 2005). The sovereign rights of states over their resources encompasses the right to grant or deny access to their genetic resources (Preamble and Article 15.1 of the CBD). Furthermore, this right is closely linked to the duty to protect and conserve biodiversity and to safeguard its sustainable use.

The CBD and Access to Genetic Resources

The basic principles of the CBD referring to access to genetic resources can be summarized as follows: Parties to the CBD are to grant facilitated access to their genetic resources (Article 15.2 of the CBD) and should receive their share of the benefits arising out of the utilization of these resources (Article 15.7 of the CBD)—the access and benefit-sharing system (ABS system).

Article 15 of the CBD provides for a legal framework to regulate access to genetic resources. Two of the most important requirements are the following: First, the CBD requires that supplier states give their prior informed consent to access. This implies that supplier states can refuse consent. Second, the parties must agree on the terms of access (mutually agreed terms), including the fair and equitable sharing of the benefits arising out of the use of genetic resources.

It should be pointed out that this legal framework was designed to give access to genetic resources for commercial purposes such as the development of pharmaceuticals (Wrigley and Chicarelli-Robinson 1997). Thus the CBD does not make a distinction between the access procedure for (basic) research and for commercial purposes. Any kind of access to genetic resources has to go through the CBD access procedure.

The CBD and Traditional Knowledge

The CBD includes in its Article 8(j) a key provision relating to indigenous and local communities. This article refers to "knowledge, innovations and practices of indigenous and local communities embodying traditional lifestyles relevant for the conservation and sustainable use of biodiversity." This is equivalent to the current understanding of the term *traditional knowledge.*

Article 8(j) of the CBD envisages three different obligations for the parties (the states) but leaves enough room for them to decide how to put this mandate into practice according to their own realities. First, it stipulates the state's obligation to "respect, preserve and maintain" such traditional knowledge. Second, Article 8(j) calls for the encouragement of "the equitable sharing of the benefits arising from the utilization of such knowledge, innovations and practices." Third, this article calls for the "approval and involvement of the holders of such knowledge, innovations and practices."

As approval and involvement are clearly mentioned in this article, this introduces a right of the respective local communities to be involved in scientific activities utilizing their knowledge in a sense that they also have the right to disapprove any use of their traditional knowledge. (See likewise Stoll and von Hahn 2004.) This article also presupposes a certain kind of prior informed consent of the respective indigenous or local communities where any science project is going to take place.

Science, Its Objectives, and the CBD: The Example of Basic Research

Scientific activities, such as biodiversity-related research, can play a substantial role for the better understanding of biodiversity and its conservation, as well as for the sustainable use of biodiversity and its components, such as genetic resources. From the point of view of the CBD, it can be said that scientific work and research cooperation are supporting tools for the fulfillment of its main objectives.

Biodiversity-related basic research (noncommercial research) involves activities in the field of taxonomy, which comprises zoology, botany, microbiology, and ecology. Basic research requires direct handling of biological resources and has specific characteristics and working methods. Biodiversity-related basic research has been carried out so far under pure research permits or has been carried out in the context of existing research cooperation between national and international research institutions (universities, scientific institutions, other public research institutions). The CBD access procedure was not designed to suit the peculiarities of basic research activities.

Biodiversity-related scientific research should be subject to a different set of rules and principles than commercially oriented research. It is possible in most cases to identify activities that are purely scientific and not for profit, although the distinction between basic research and applied and more commercially oriented research is sometimes blurred.

The CBD and Global Environmental Change (GEC) Science

The CBD's regulations and its working programs, decisions, and guidelines also affect the GEC science research portfolio, including ecosystem assessments, environmental and land use research, and so on. By touching the same subject matter—biodiversity and its different components—the CBD's regulations are having an impact on the GEC science research and therefore should be taken into account at the outset of defining project structures and aims.

Many projects of basic GEC research are often linked to biodiversity research and involve components of biodiversity such as ecosystems, species, and genes. Some GEC-driven projects might involve access to genetic resources. Such GEC projects could be connected to CBD regulations. It is a challenge and also a responsibility of the respective projects to be aware of and identify these possible connections to the CBD regulations and to observe these rules, for example, in the above-mentioned involvement of traditional knowledge.

In the context of traditional knowledge, the CBD requires at least the consultation of indigenous or local communities when GEC-driven projects involve the use of traditional knowledge. Therefore from the outset, project principal investigators should take into account the necessity of obtaining prior informed consent of the relevant indigenous or local communities in the project area. This should encompass in a transparent way the aims, structure, and different phases of the project; the common understanding of benefits; the potential risks; and how to share the mutually agreed benefits arising out of the access activities.

Besides the above-discussed relations of the GEC-driven projects and the CBD access procedures to genetic resources and traditional knowledge, it should be pointed out that the links between the CBD implementation and the GEC science portfolio are constantly increasing. For example, Decision VII/3 concerning agriculture biodiversity invited government and funding agencies, among others, to conduct case studies based on socioeconomic and ecological analyses of different land use management options. This kind of case study could be a GEC science project. The main message is therefore that GEC projects should observe CBD legal requirements and, at the same time, may count as contributions to the convention.

The Implementation of the CBD's ABS System in Latin America and Its Impact on Biodiversity-Related Basic Research

The manner of implementation of the established ABS system of the Convention on Biological Diversity is at the states' discretion. Signatory parties are requested to incorporate the main principles named in Article 15 of the CBD in their national legislation. Currently states have full discretion to introduce and tighten administrative requirements, obligations, and benefit-sharing duties.

The impact on science activities is growing as individual countries implement CBD work programs, applying CBD decisions and guidelines. This is becoming palpable in the case of scientific research related to biodiversity and its components, such as genetic resources (Feit et al. 2005). By requiring specific procedures for any type of activities concerning genetic resources, the new developments in the CBD process may make access to genetic material for basic research increasingly difficult, as researchers' activities fall under new requirements originally designed to regulate access to biological resources for commercial purposes.

The introduction of an international regime on access and benefit-sharing, negotiated during the Fourth Session of the Working Group on Access and Benefit-Sharing of the CBD in Granada, Spain (January 30 to February 3, 2006) and the ordinary meeting of the Conference of the Parties to the CBD in March 2006, may further complicate the access procedure.

A closer analysis of the existing international regulatory system implementing the ABS system under the CBD shows that access applicants for basic research purposes are confronted with unclear procedures and very restrictive legislation in many countries. In several cases, domestic regulatory frameworks show contradictory stipulations or simply a lack of necessary legislation.

The following major consequences for basic research can be identified:

- increased bureaucratic hurdles for the exchange of biodiversity-related information among scientists and research foundations
- potential administrative obstacles in access legislation
- adoption of more strict legal requirements for basic research
- tightening of export conditions concerning biological material for research
- imposed necessity to change or adjust existing cooperation models and research practice
- danger of being measured by the same standards as research for commercial applications (in particular, concerning benefit-sharing expectations)
- increasing difficulties in building up trust among all stakeholders

These consequences apply to any project related to any component of biodiversity and therefore also to GEC science research (Feit et al. 2005).

The Case of Ecuador

In the context of the Andean Community of Nations (CAN), the relevant legal framework for bioprospecting projects is constituted by the CBD, the common regime on access to genetic resources established by Decision 391, and domestic legislation complementing Decision 391 (CAN 1996).

Decision 391 contains provisions regulating the access procedure for genetic resources. The procedure does not make a distinction between bioprospecting projects for research purposes and projects with a commercialization or industrial aim. Contrary to the CBD, Decision 391 provides a definition for the term *access*: the obtaining and use of genetic resources conserved *in situ* and *ex situ*, of their by-products, and, if applicable, of their intangible components, for purposes of research, biological prospecting, conservation, industrial application, and commercial use, among other things. As a consequence, in accordance with Decision 391, all types of access activities are subject to the same set of rules (Article 1 of Decision 391).

Unlike the CBD, Decision 391 includes in its scope the by-products of genetic resources. According to Article 1, a *by-product* is "a molecule, a combination or mixture of natural molecules, including crude extracts of live or dead organisms of biological origin that come from the metabolism of living beings." Consequently the access mechanism is to be applied also to by-products such as plant extracts. Here the scope is broadened so, for example, in the case of plant extracts, an access procedure is to be followed.

According to Decision 391, the applicant has to provide, as part of the application, information about the applicant, the access activity applied for, the area of research, and so on. The referential model for the application, approved through Resolution 414, includes the attachment of a declaration of consent from the supplier of genetic resources or traditional knowledge (*intangible component* in the decision parlance). However, the content of the declaration has not been specified. Due to its referential character, the application model could eventually be modified by CAN member countries through domestic regulatory activities.

As part of the access contract, Decision 391 requires applicants to incorporate an annex containing arrangements for sharing of benefits arising out of the use of genetic resources when relevant traditional knowledge of indigenous communities is involved.

Although the Andean norms are obligatory and require immediate fulfillment by the CAN member countries, Decision 391 allows members to enact complementary domestic legislation (Article 7 of Decision 391). Ecuador is currently drafting a national regulation aimed at the establishment of complementary provisions for the application of Decision 391—the *Reglamento* (forthcoming).

The Ecuadorian *Reglamento* draft proposal further elaborates on the access procedures of genetic resources. To obtain access authorization for genetic resources, the applicant should present a request in line with Decision 391. However, the *Reglamento* introduces a new requirement when indigenous or local communities are involved as providers of associated traditional knowledge (intangible component) of genetic resources for which access is requested. A proposal for a consultation process with the communities shall be included along with the request (Article 25 of the *Reglamento*). Little guidance for the proposal is given in the *Reglamento*. It stipulates, first, that the proposal should respect indigenous communities' traditional organizational practices, mechanisms, and forms and, second, that the proposal must include the goals, phases, mechanisms, activities, and methodologies to proceed with the consultation of the indigenous communities.

However, it does not indicate the specific subject matter to be consulted, nor does it provide guidance about the aim of the consultation process, leaving it to the applicants to organize the entire process at their own risk and in good faith.

The Ecuadorian draft proposal has met with criticism from various indigenous organizations in Ecuador. They focus on the non–legally binding nature of the results of the consultation process and on the lack of clear benefit-sharing regulations. The above explanations show that applicants requesting access to biological resources in Ecuador currently face uncertainties concerning the access procedure, independent of the type of access and particularly when traditional knowledge is involved.

The German project ProBenefit is an example of this situation (http://www.pro benefit.de). ProBenefit is a bioprospecting project under the CBD with the goal to work out a model procedure and agreement for a fair and transparent sharing of benefits from the sustainable use of biodiversity in the Amazon lowland of Ecuador. ProBenefit aims at developing an access agreement in transparent negotiations with local communities in Ecuador, relevant nongovernmental organizations, the Ecuadorian authorities, and a medium-sized pharmaceutical company from Germany. It is funded by Germany's Federal Ministry for Education and Research under the biodiversity research program BioTeam.

The project has been confronted with the following challenges:

- Due to the lack of regulatory certainty, there has so far not been any successful application for a bioprospecting project. Since 2003 ProBenefit has been maintaining intense dialogue with the relevant authorities to find a way to begin the access procedure officially.
- The draft *Reglamento* leaves in place several uncertainties regarding, among other things, the legal nature of the consultation process with relevant indigenous communities and the basic elements of benefit-sharing arrangements.
- This situation has prevented ProBenefit from beginning any kind of research in Ecuador.
- Concerning indigenous communities, the project is facing the challenge of designing, together with the relevant communities and competent national authorities, the consultation process as proposed by the draft *Reglamento*.

Experiences Gained by Analyzing ABS System Implementation in Latin America

Latin American countries have shown a very strong interest in developing policies as well as legal frameworks to regulate access to genetic material. The member states of CAN are implementing their common regime on access, which entered into force in 1996.

The positive aspect is that these national implementations of the ABS system under the CBD comprise the principles of prior informed consent, mutually agreed terms, and benefit sharing. In addition they highlight in different ways the importance of traditional knowledge for the conservation and sustainable use of biodiversity. However, they show different tendencies in their access procedures. Some member states (Venezuela, Colombia) apply directly the procedures of Decision 391, while others (Bolivia, Peru) have enacted and one (Ecuador) is in the process of enacting complementary regulations.

Unfortunately, many of these implementations do not really contribute to the provision of a clear, unbureaucratic, and effective procedure to support the enforcement of the CBD provisions. As a side effect, they unnecessarily hinder scientific work.

The regulations also lack a clear distinction between different types of access procedures. Therefore, pure basic research access activities especially are still in danger of being measured by the same standards as research aiming at commercial application. Uncertainties related to which government bodies are authorized to grant access often result in confusion and organizational difficulties in planning, arranging, and undertaking access activities.

Experiences like those of ProBenefit in Ecuador have shown that it is very important to analyze national ABS legislation and research permit procedures before beginning with the official access procedure. The ProBenefit experience also shows that maintaining a constant dialogue with the national authorities is very important in order to avoid misunderstandings; to achieve transparency, legitimacy, and trust concerning the access activities; and to foresee and prevent future difficulties over the whole period of activities. Keeping in regular contact with competent national authorities and the relevant stakeholders is an important working method to keep projects moving forward.

From the point of view of basic biological research—often the basis for bioprospecting projects like ProBenefit—it can be said that scientists have to deal with a new legal framework that, on the one hand, recognizes the importance of this kind of research but, on the other hand, does not take into account its special status and needs.

A crucial issue in regard to the existing policies and laws is that of finding appropriate mechanisms for a distinction between the various types of access activities (basic research, applied science, research directed to commercialization process, etc.) in order to adapt the regulations to their peculiarities, specific objectives, and needs. The existing ABS system policies and laws such as Decision 391 do not contribute to this goal.

Twelve years after the CBD entered into force, the database of its secretariat comprises national legislations of 29 countries. These national legislative measures must be observed by any project or scientist doing research on and working with biodiversity. However, the majority of users of genetic resources are insufficiently informed about the CBD and the respective legal framework regarding the access and use of genetic resources and associated traditional knowledge. Many stakeholders (e.g., researchers, project leaders, research institutes, sponsors, technology transfer units, funding agencies) do not really know whether CBD regulations are relevant to them or not.

To avoid and prevent negative effects of existing or new legal developments on their funded projects, funding agencies could analyze the creation of their own ABS policies, providing training courses to keep their applicants for funding, as well as the project evaluators, informed about the new legal developments related to the use of biodiversity and sensitive topics such as traditional knowledge. These ABS policies could encompass the development of codes of conduct or ethical guidelines on ABS or new funding guidelines, to safeguard that scientists in charge of the biodiversity-related projects observe and fulfill relevant international and domestic legal provisions.

These instruments could be an effective tool to increase trust among the local project partners, and they could be a contribution toward doing science in a legitimate and transparent way under this international treaty on the rights of local and indigenous communities. As an example, some user countries such as Denmark or Sweden require the fulfillment of access and benefit-sharing requirements as a prerequisite for public funding of research and development projects related to biodiversity.

These new developments represent new challenges for the scientific community in the conception and formulation, as well as in the management, of biodiversity-related research projects, often generating new costs. Funding agencies should take these compliance issues into account and also consider their impact on future project budgets.

Literature Cited

CAN. 1996. *Decisión 391: Régimen Común sobre Acceso a los Recursos Genéticos. Publicada en el Suplemento al Registro Oficial No. 5 de 16 de agosto de 1996.* Comunidad Andina de Naciones.

CBD. 2005. *Handbook of the Convention on Biological Diversity* (including its Cartagena Protocol on Biosafety), 3rd ed. Montreal: Convention on Biological Diversity. (p. 93) (http://www.biodiv.org/convention/convention.shtml)

Feit, U., M. Driesch von den, and W. Lobin, eds. 2005. *Access and benefit sharing of genetic resources: Ways and means for facilitating biodiversity research and conservation while safeguarding ABS provisions.* Skripten 163. Bonn: Federal Agency for Nature Conservation (BfN).

Gehl Sampath, P. 2005. *Regulating bioprospecting: Institutions for drug research, access and benefit-sharing.* UN University Press. (www.intech.unu.edu/publications/)

Reglamento a la Decisión 391 del Acuerdo de Cartagena Relativa al Régimen Común sobre Acceso a los Recursos Genéticos.

Stoianoff, N. 2004. *Accessing biological resources: Complying with the Convention on Biological Diversity.* The Hague: Kluwer Law International. (p. xv)

Stoll, P.-T. 2004. Genetische Ressourcen, Zugang und Vorteilsteilhabe. In Wolff, N., and W. Köck, eds. *10 Jahre Übereinkommen über die biologische Vielfalt – Eine Zwischenbilanz.* Baden-Baden: Nomos Verlagsgesellschaft. (pp. 73–88)

Stoll, P.-T., and A. von Hahn. 2004. Indigenous peoples, indigenous knowledge and indigenous resources in international law. In von Lewinski, S., ed. *Indigenous heritage and intellectual property: Genetic resources, traditional knowledge and folklore.* The Hague: Kluwer Law International. (p. 5–47)

Wrigley, S. K., and M. I. Chicarelli-Robinson. 1997. Natural products research and pharmaceuticals in the 1990s. *Annual Reports in Medicinal Chemistry* 32:285.

PART III
Examples of the Science-Policy Interface

11

Integrating Environmental and Social Agendas: The Experience of the Amazonian Networks LBA and GEOMA

Peter M. de Toledo, Ima C. G. Vieira, Gilberto Câmara, and Carlos A. Nobre

Uncovering ecological complexity under a changing environment requires multidisciplinary and interdisciplinary approaches for the development of future scenarios that are crucial for planning human welfare while maintaining environmental stability. Biological diversity worldwide has experienced a marked decline in recent years due to land cover conversion, climate change, contamination, overexploitation of natural resources, and introduction of exotic species (PNUMA 2002). These intertwined factors constitute a set of human-induced drivers that affect most of Earth's ecosystems. Human needs for food and fiber account for a significant portion of the primary productivity of the planet. Emission of greenhouse gases arising from fossil fuels and air pollution by a large array of pollutants contribute further negative impacts on natural landscapes and on humans. This unbalanced intense modification of worldwide ecological conditions induces feedbacks from natural systems, especially of old-growth forests that are slowly being unraveled by the scientific community (Nobre and Nobre 2002; Lewis et al. 2004). Only 17 percent of the pristine landscapes still remains untouched, mostly in tropical regions and in rain forest domains (Dirzo 2001). The tropical belt (30°N–30°S) accounts for 75 percent of the world's population, 20 percent of food production, and 80 percent of the world's births. Understanding the environmental conditions and diversity patterns on such a large domain of terrestrial ecosystems is a difficult task for the scientific community. Human impact and its social, cultural, and economic context must be taken into account when studying modern ecosystems. Results of such endeavors will contribute to our knowledge about the planet's functioning and the global change humans are imposing on it.

In the coming decades, Brazil faces daunting challenges in the management of its highly diverse and threatened ecological systems under the paradigm of sustainable development. Complex problems, ranging from environmental impacts of the ever-growing and fast-paced human expansion over Brazilian natural biomes to the urban concentrations that affect life quality, will require constant improvement in the national capacity to model and simulate scenarios of public policies with explicitly defined spatial and temporal dimensions. Due to disparities in the geographical scale, ecosystem diversity, and regional socioeconomic conditions, past and current analyses of the overall development process have not followed an integrated and systematic approach. Until recently only a few governmental programs had been formulated that combine economical factors, ecological issues, demographic characteristics, and climatic patterns. Consequently patterns of public policy implementation in Brazil are usually still governed by a simplistic disciplinary perspective. This context leads to a pattern where each individual sector of the government makes decisions under a limited or narrow perception of its impacts and environmental consequences. This is particularly evident in the northern region dominated by the Amazonian biome.

In the Brazilian Amazon, two major research programs hosted by the Brazilian Ministry of Science and Technology have been developed in order to produce science outputs concerning the Amazon's rapid changes in land cover and climatic variability. The first effort was the Large-Scale Biosphere-Atmosphere Experiment in Amazonia (LBA), which is the largest international collaborative research project ever organized to study tropical ecosystems. Its major objective is understanding the functioning of the Amazon as a regional entity with respect to natural cycles and understanding how the physical, chemical, and biological functioning of ecosystems are changing in response to land use and climate changes. More recently the Environmental Modelling Network of the Amazon (GEOMA network) was created to articulate scientific expertise and consolidate information and knowledge toward public policy. This chapter's main objective is to demonstrate the scientific importance of both of these ongoing programs and to show how they effectively collaborate to build an important conduit for communicating science to policy makers in Amazonia.

Amazonia as a Valuable Lesson for Scientists and Policy Makers

Amazonia, with its formidable landscape variability, environmental peculiarities, and important socioeconomic features, has attracted the attention of the academic world (Becker 2005; Wright 2005). In particular, regulation and the region's wealth of biodiversity have drawn the attention of natural and social scientists working to understand ecosystem functioning and sustainability in the face of rapid change (Fearnside 2003; Bermingham et al. 2005). There is an ongoing trend of networking in the scientific community aiming to understand the Amazon forest from multidisciplinary and interdisciplinary perspectives with the idea that such a multifaceted biome can be understood only by combining different fields and views of the academic world and by a cooperative and complementary effort. Research programs in Amazonia must consider all major physical setting and biotic landscape components and characteristics through the collection of raw data on both spatial and temporal scales. The lack of basic information is still the major problem concerning the understanding of this complex natural system. The main conclusion is that human interference on the Amazonian environment without a minimum of information on ecosystem functioning and the potential feedbacks resulting from land conversion will rarely be sustainable. Some of the important questions are these: 1) How can the interactions between nature and society be better incorporated in efforts that integrate natural environment, human development, and sustainability? 2) How do large-scale trends of global changes act on the environment to affect development and quality of life of local populations? 3) How can independent activities such as research, planning, monitoring, and decision making interact?

Science and Public Policy I: Lessons from the LBA Experiment

The LBA has overarching scientific goals, namely 1) to understand Amazonia as a regional entity with respect to the natural cycles of water, greenhouse and other important gases and aerosols, and nutrients; 2) to understand how land use and climate changes affect the physical, chemical, and biological functioning of ecosystems; and 3) to understand the impacts of those changes on the sustainability of Amazonia and their influence on the global climate. LBA has become one of the most interdisciplinary and integrated regional studies ever carried out in the world. Since its planning stage, which took a long period (1993–1998), LBA formed as a large network of researchers and scientific institutions from many parts of the world, both developing and developed. The long planning phase allowed for the slow building up of a broad-based network and also permitted a thorough discussion of its scientific goals. The result was a formidable network involving over 1,000 researchers and students and more than 200 institutions from Brazil, all of the other Amazonian countries, the United States, and 9 European countries. LBA has become the flagship of global environmental change programs, such as the International Geosphere-Biosphere Programme and the World Climate Research Programme, as a model of "integrate regional studies," and is a role model for similar regional studies in other parts of the planet (e.g., the Monsoon Asia Integrated Regional Study). Seven years after its start, LBA has achieved most of its scientific goals, in terms of new knowledge on Amazonian ecosystems and environments and of the impacts of current changes, as demonstrated by over 600 peer-reviewed publications by the middle of 2006. LBA has been instrumental in building scientific capacity in Amazonia as a result of its successful training of over 300 masters and doctoral students, mostly from the region. This is one long-term legacy of LBA to Amazonia that is already producing results, such as the establishment of novel postgraduate and graduate programs in many Amazonian universities.

However, LBA was less effective in delivering more definite results to contribute to the development of a model for sustainability of Amazonia. That deficiency highlights a serious shortcoming of global environmental research, which is its separation from the scientific basis of development policies. Contrary to early expectations in LBA, translating new scientific results into sustainable development policies for Amazonia proved to be quite a challenge (Lahsen and Nobre forthcoming). Beyond capacity building, LBA's mission was to provide the scientific basis to safeguard basic ecological processes. With that focus, it is perhaps not surprising that it performed weakly in social, political, and economic implications of the science, with few exceptions. Providing the foundations for sustainability science for Amazonia is still an unmet goal for Amazonian science in general. A new and sustainable development paradigm for the region must encompass many more dimensions than the basic knowledge on ecosystems and how they respond to change. A new development paradigm must also take into account the needed, and not yet available, appropriate technologies to tap the wealth hidden in the forest's biodiversity.

We here present two illustrations of LBA results and their potential implications for policy making, one related to climate change and the other related to local agricultural development.

The Amazonian Carbon Balance and the Role of Avoided Deforestation for Mitigating Climate Change

Understanding the carbon cycle of Amazonian ecosystems and how it is affected by deforestation, fire, forest fragmentation, and so on has been one of the scientific goals of LBA. With the results of a number of well-conducted field, remote sensing, and modeling studies on the carbon cycle in LBA, it became evident that avoiding deforestation should be given high priority as a practical measure to reduce emissions of greenhouse gases for two commanding reasons: on one hand, the forest stores about 30 to 40 times more carbon than the replacement vegetation and, on the other hand, undisturbed forests may play an important role as additional sinks of atmospheric carbon. (See a review of Amazonian carbon cycle studies in Ometto et al. 2005.)

The wealth of scientific evidence was an important factor in bringing this issue to the forefront of climate change discussions, including underpinning the proposal of mechanisms to compensate tropical countries that reduce deforestation (Santilli et al. 2005). The 11th session of the Conference of the Parties to the Climate Change Convention (COP11) finally accepted the decision to include in future discussions *avoided* deforestation as a valid means to reduce emissions of greenhouse gases. The COP11 agenda stated, "The COP invited Parties and accredited observers to submit to the secretariat, by 31 March 2006, their views on issues relating to reducing emissions from deforestation in developing countries, focusing on relevant scientific, technical and methodological issues, and the exchange of relevant information and experiences, including policy approaches and positive incentives." The issue was further discussed at COP11 when the Brazilian delegation, led by Brazilian Minister of Environment Marina Silva, actually proposed the creation of a voluntary fund to compensate tropical countries that demonstrated reductions of deforestation. It is quite possible that a mechanism of the sort proposed by Santilli et al. (2005) and denominated "compensate reductions" will be established in the next few years to compensate developing countries that avoid emissions by means of decreasing deforestation. The practical implementation of such a mechanism will raise scientific, technical, and methodological issues. Many of these already have been dealt with by LBA and GEOMA. One example is the use of sophisticated remote sensing tools to assess and quantify vegetation cover changes in Amazonia and their implications for the carbon balance.

Projecting the impact of future climate change on vegetation in Amazonia highlights another LBA result with importance to biodiversity conservation policies, as well as land use, water resources, and agricultural development policies. Global climate change, acting in synergy with regional deforestation, may increase fire frequency and forest fragmentation and result in the tendency of "savannization" of large portions of Amazonia (Oyama and Nobre 2003; Nobre et al. 2005). Therefore public policies relating to biological conservation, including of hot spots for conservation (ecological corridors, among others), must take into account the possibility of significant changes of Amazonian biomes due to climate change alone. Losing precious biological diversity is again one more example of the likelihood that developing nations will bear the brunt of climate change even with minimal responsibility for causing the problem.

Studies on the Ecological and Social Importance of Secondary Vegetation

According to Vieira et al. (2003), around 40 percent of the agricultural area in the Brazilian state of Pará is occupied by fallow vegetation and only 8 percent to 11 percent of the total rural area is being occupied by cropland. In contrast to this enormous extension of fallow areas, around 12 percent of the land is covered by residual primary forest, most of it in riparian strips, after a century of settlement and agricultural land use in that area. Despite the importance of secondary vegetation for settlers, little is known about how these farmers utilize secondary vegetation in their farming systems. Existing models suggest that vegetative fallow provides natural re-fertilization and soil protection, but the work remains abstract, with little empirical grounding in the knowledge base of ecology. This shortcoming inhibits models of carbon flux that are based on human behavior, which realize that regrowth vegetation sequesters carbon and that it is a function of farming activity.

Secondary vegetation is a type of land cover, but it is also a conceptual focal point that unites information from the ecological, physical, and social sciences. As such, its study is inherently interdisciplinary in focus, toward the understanding of the dynamics and management of anthropogenic vegetation in these landscapes (Wu 2006). Ecologists describe the structure of secondary vegetation and its rate of biomass accumulation, and they measure the area extent of succession and the environmental conditions, including those of human origin, that impact its trajectory of change (Uhl 1987; Vieira et al. 2003). Social scientists explain why people find it advantageous to manage fallows, to slash their regrowth vegetation, and to abandon fields to natural processes of nutrient recovery (Moran et al. 1994). For the purposes of modeling, LBA and other related projects are developing a conceptual framework that allows for integration across disciplines. The results show that forest regrowth in the Brazilian Amazon constitutes a significant component of the regional carbon balance (Houghton et al. 2000; Ometto et al. 2005). It also serves to restore hydrological functions, plant and animal habitats, and landscape connectivity (Vieira et al. 1996; Metzger 2003). Even after many years of moderate land use, abandoned areas can recover 30 percent to 35 percent of floristic composition (compared with primary forest) and 60 percent of biomass after 70 years of succession (Vieira et al. 1996; Vieira et al. 2003). Such information has been used to formulate public agriculture planning, especially in those regions prone to slash-and-burn agriculture by small farmers.

Science and Public Policy II: Lessons from the GEOMA Network

The main objective of the GEOMA network is to develop models to evaluate and to provide scenarios of sustainability for Amazonia under different natural and human science aspects in order to make useful contribution to public policy. The basic requirement for formulating such scenarios is the integration of socioeconomic, environmental, demographic, and climatic parameters in a strong interdisciplinary approach. Therefore defining lines of multidisciplinary and interdisciplinary research constitutes the main task of the network. Divided into thematic areas, each individual disciplinary component has the task of formulating specific questions and providing solutions that must be applicable to development and/or environmental issues. Most research topics are based upon relevant aspects of Brazilian public policy.

The major areas of interest of the GEOMA network are the following:

- Analyze the relationship of land use and land cover change to the productive systems, considering agriculture, livestock, the structure of the agrarian system, and the conditions for quality of life in the productive sector.
- Develop models to create protection areas for biological conservation in Amazonia.
- Develop, test, and integrate models for aquatic ecosystems.
- Investigate and model demographics in Amazonia, in particular the component of socio-spatial mobility of the human population.
- Conduct studies of regional economy, using logistic models and regional development.

- Build a robust database.
- Develop integrated models at multiple scales that incorporate different dimensions of sustainability in Amazonia, such as human dynamics, biodiversity, land use and land cover change, and climatic and hydrological parameters.

Despite its recent conception, the scientific agenda of GEOMA has been rapidly accepted and understood by decision and policy makers at the management level in the federal government of Brazil. The construction of the scientific agenda is consistently elaborated according to critical social and environmental issues, resulting in the short-term problem-solving projects customized for rapid information delivery to decision makers. This strategy has gained rapid dissemination to politicians and governmental offices by means of technical reports. As such, it has been successful in the dialogue with policy makers by presenting results related to GEOMA's capacity of developing comprehensive studies, monitoring land cover and land use change in Amazonia. Some of its topics, especially those dealing with short-term problem-solving of social and land conversion issues, found great appeal in public environmental sectors. The network is inserted in a federal plan and is included in the official inter-ministerial program responsible for reducing deforestation and minimizing its environmental and social impacts in the Amazon region. As part of the action plan to control and prevent further erosion of the primary forests, there is an official commitment by the GEOMA network to present models to decision makers as well as to publish interdisciplinary papers under a multi-institutional framework.

Preliminary results are examples of such interaction. GEOMA scientists have organized a critical evaluation of the effectiveness of the Brazilian official conservation plan, ARPA (Programa de Áreas Protegidas da Amazônia), regarding the location and establishment of protected primary forest areas. In this case, modeling permitted the identification of crucial ecoregions that had not previously been considered for protection. These areas house unique natural communities and biological entities that are important biodiversity reservoirs.

In addition, an in-depth analysis of the main drivers that are causing the extensive forest conversion in the southeastern portion of the Amazon, especially those linked to agricultural and livestock economies and to illegal activities, has produced an interesting outline of the major hot spots of deforestation there (Câmara et al. 2005; Escada et al. 2005). This problem-solving approach provides fundamental information to policy makers who are responsible for infrastructure planning in the region.

Final Considerations

One of the contributions of the scientific community to the global change agenda is the implementation of long-term research combining information and knowledge under the concept of "sustainable territories," as defined by Vieira et al. (2005): "a mosaic of complementary land use segments of a specific territory managed under an integrated design, which enables the conservation of the biodiversity, as well as the maintenance of the dynamic of the ecological processes and allowing socioeconomical development." A prior requirement is to have defined the *territorial unit of analysis*, an area that cannot be smaller than either a municipality or a hydrographic basin. Different elements of the local society, such as different governmental levels, social organizations, and scientific institutions, must be engaged in setting goals. Reliable data should be acquired, organized, and integrated in a consistent and cohesive manner, enabling fast retrieval and assessment of information. Dissemination of results through outreach programs and education is a priority. Engaging local institutions is fundamental to create long-term commitment of public organizations. This strategy will provide positive feedback and strengthen the regional scientific system. A good option for obtaining fast and comparable results is the establishment of several pilot studies throughout the region.

As an example of such an approach, one can cite the study of food production and conservation. It is generally recognized that there are at least three different dimensions to food security research. The macroscale food supply dimension focuses on the extent to which future food products might keep pace with population growth and thereby ensure a reliable supply of basic food products. The mesoscale perspective is concerned with the broad access to and availability of food. The microscale food security dimension recognizes that many of the impediments to food security are tied to fundamental human needs and rights. Opportunities for employment and the capacity to adapt to environmental and social economical pressures are viewed as prerequisites to achieving food security. In this context, food security in Amazonia cannot be separated from the environmental agenda. There is an urgent need to bring the food security approach to Amazonian studies, while also integrating the global change and socioeconomic agendas. Food security research must focus on land conservation and strengthening of agricultural research, and it may be part of studies of global environmental change that may transform societies at all scales. If this is realized, we improve our understanding of differential vulnerabilities of Amazonia in areas where conservation goals are very important, such as in the states of Amapá and Amazonas. In both states, 80 percent of the poultry, meat, and manioc flour consumed are imported from the state of Pará, which already has reached the cap of 20 percent of legally permitted deforestation. The conservation pressure on specific regions of Amazonia depends also on the organization of food production networks for the people who live in the region. Therefore the conservation of Amazonia must consider human population growth and basic needs and the reorganization of food production areas. For the preservation of certain areas in Amazonia, there is an urgent need to optimize technological practices in those regions already altered.

While solutions to sustainability problems in the Amazon in some cases may be found through technology, the problems are responses to national and global level economic structures that perpetuate poverty, ignorance, and unsustainable, short-sighted extractive approaches to natural resource management. To truly understand and address environmental degradation in the Amazon, one must also strengthen the understanding and recognition of the connections between sustainability problems and global and regional structures of power and inequality, including the impact of capitalism and liberal globalization on environmental practices, standards, and policies. The LBA and GEOMA projects and their associated networks have been addressing such questions over the past years, showing that scientists can communicate and effect changes in policies toward environmental and climate change.

Literature Cited

Baar, R., M. R.Cordeiro, M. Denich, and H. Folster. 2004. Floristic inventory of secondary vegetation in agricultural systems of East-Amazonia. *Biodiversity and Conservation* 13:201–528.

Becker, B. K. 2005. Amazônia. Geopolítica na virada do III milênio. Rio de Janeiro: Garamond. (168 pp.)

Bermingham, E., C. W. Dick, and C. Moritz. 2005. *Tropical rainforests: Past, present, and future.* Chicago: University of Chicago Press. (745 pp.)

Câmara, G., A. P. D. Aguiar, M. I. S. Escada, S. Amaral, T. Carneiro, A. M. Monteiro, R. A. Santos, Jr., I. C. G. Vieira, and B. Becker. 2005. Amazonian deforestation model. *Science* 307:1043–4

Dirzo, R. 2001. Tropical forests. In: Chapin, F. S., O. E. Sala, E. Huber-Sannwald, eds. *Global biodiversity in a changing environment: Scenarios for the 21st century.* Springer. (Ecological Studies Series, vol. 152, pp. 251–76)

Escada, M. I. S., I. C. G. Vieira, S. Amaral, R. A. Santos, Jr., J. B. Veiga, A. P. D. Aguiar, I. Veiga, M. Oliveira, J. Gavina, A. Carneiro Filho, F. Fearnside, A. Venturieri, F. Carrielo, M. Thales, T. S. Carneiro, A. M. Monteiro, and G. Câmara. 2005. Padrões e Processos de Ocupação nas Novas Fronteiras da Amazônia: Apropriação Fundiária e Uso da Terra no Xingu/Iriri. *Estudos Avançados* 54:9–23.

Fearnside, P. M. 2003. *A floresta amazônica nas mudanças globais.* Manaus: Instituto Nacional de Pesquisas da Amazônia. (134 pp.)

Houghton, R. A., D. L. Skole, C. A. Nobre, J. L. Hackler, K. T. Lawrence, and W. H. Chomentowski. 2000. Annual fluxes of carbon from deforestation and regrowth in the Brazilian Amazon. *Nature* 403(6767):301–4.

Lahsen, M. and C. A. Nobre. Forthcoming. Challenges of connecting international science and local level sustainability efforts: The case of the LBA. *Environmental Science & Policy.*

Lewis, S. L., O. L. Phillips, T. R. Baker, J. Lloyd, Y. Mahli, S. Almeida, N. Higuchi, W. F. Laurance, D. A. Neill, J. N. M. Silva, J. Terbough, A. T. Lazama, R. V. Martinez, S. Brown, J. Chave, C. Kuelber, P. N. Vargas, and B. Vincetti. 2004. Concerted changes in tropical forest structure and dynamic evidence from 50 South American plots. Philosophical Transactions of the Royal Society of London. *Biological Sciences* 359:421–36.

Metzger, J. P. 2003 Effects of slash-and-burn fallow periods on landscape structure. *Environmental Conservation* 30:325–33.

Moran, E.F., E. Brondizio, P. Mausel, and Y. Wu. 1994. Integrating Amazonian vegetation, land-use and satellite data. *Bioscience* 44:329–38.

Nepstad, D. C., P. Moutinho, and D. Markewitz. 2001. The recovery of biomass, nutrient stocks, and deep-soil functions in secondary forests. In McClain, M. E., R. L. Victoria, and J. E. Richey, eds. *Biogeochemistry of the Amazon Basin.* New York: Oxford University Press. (pp. 139–155)

Nobre, C. A., and A. D. Nobre. 2002. O balanço de carbono da Amazônia brasileira. *Estudos Avançados* 45:81–90.

Nobre, C. A., E. D. Assad, and M. D. Oyama. 2005. O Impacto do Aquecimento Global nos Ecossistemas da Amazônia e na Agricultura. *Scientific American Brasil* 5:70–5.

Ometto, J. P. H. B. , A. D. Nobre, H. R. Rocha da, P. Artaxo, and L. A. Martinelli. 2005. Amazonia and the modern carbon cycle: Lessons learned. *Oecologia* DOI: 101007/s00442-005-0034-3.

Oyama, M. D., and C. A. Nobre. 2003. A new climate-vegetation equilibrium state for tropical South America. *Geophysical Research Letters* 30(23):2199.

PNUMA. 2002. *Perspectivas del Médio Ambiente Mundial 2002 GEO-3.* Programa de las Naciones Unidas para el Medio Ambiente, UNEP. (446 pp.)

Santilli, M., P. Moutinho, S. Schwartzman, D. Nepstad, C. A. Nobre, L. Curran. 2005. Tropical deforestation and the Kyoto Protocol. *Climatic Change* 71:267–76.

Uhl, C. 1987. Factors controlling succession following slash and burn agriculture in Amazonia. *Journal of Ecology* 75:377–407.

Vieira, I. C. G., A. Almeida, E. Davidson, T. Stone, C. R. Carvalho, and J. B. Guerrero. 2003. Classifying successional forests using Landsat spectral properties and ecological characteristics in eastern Amazônia. *Remote Sensing of Environment* 87(4):470–81.

Vieira, I. C. G., R. Salomão, N. Rosa, D. Nepstad, and J. Roma. 1996. Renascimento da floresta no rastro da agricultura. *Ciência Hoje* 20:38–45.

Vieira, I. C. G., J. M. C. Silva, and P. M. Toledo. 2005. Estratégias para evitar a perda de biodiversidade na Amazônia. *Estudos Avançados* 54:153–64.

Wright, S. J. 2005. Tropical forests in a changing environment. *Trends in Ecology and Evolution* 20(10):553–60.

Wu, J. 2006. Landscape ecology, cross-disciplinary, and sustainability science. *Lancscape Ecology* 21:1–4.

12

Assessment of Present, Past, and Future Climate Variability in the Americas from Treeline Environments

Brian H. Luckman

The Inter-American Institute for Global Change Research (IAI) framework provided a unique opportunity to develop an integrated, interdisciplinary, and international project to address climate variability along a global transect of the Americas. The primary goals of the Collaborative Research Network (CRN) on "the assessment of present, past and future climate variability in the Americas from treeline environments" were the following:

- Develop a network of tree-ring chronologies from climatically sensitive treeline sites in the western American cordillera.
- Use these data to reconstruct and compare regional climate variability along a transect from Alaska to Tierra del Fuego.
- Define the characteristic modes of that variability at interannual, decadal, and longer timescales.
- Accelerate the development and use of dendrochronology for tropical mountain tree species in the Americas as a means of addressing the issues of climate variability and global change.

The project also sought to enhance and expand collaboration and training in dendrochronology and paleoenvironmental science within Latin America.

The 18 cooperating scientists of the CRN came from 13 institutions in Argentina, Bolivia, Canada, Chile, Mexico, Peru, and the United States. The group includes scientists from 10 universities, three government agencies, and one research laboratory, who collectively are associated with 12 tree-ring laboratories of different sizes, 3 of which were created by the project. Despite the dendrochronological focus, these co-principal investigators (Co-PIs) brought diverse academic backgrounds to the project, with degrees in agricultural engineering, anthropology, archeology, botany, climatology, earth sciences, ecology, forestry, geography, geomorphology, hydrology, physics, and watershed management, thus giving the CRN a truly interdisciplinary flavor.

Major Scientific Achievements

During the CRN project, over 300 new tree-ring chronologies were collected throughout the cordillera at sites between 66°56'N (*Picea glauca*, Yukon) and 55°50'S (*Nothofagus pumilio*, Cape Horn). Elevations ranged from 4,670 m (*Polylepis tarapacana*, Soniquera, Bolivia) to sea level in southern Chile (*Pilgerodendron uviferum*). The collections included six major single-species networks and chronologies from eight species new to dendrochronology. Although most chronologies are between 200 and 400 years, the collections include a 5,666-year chronology from *Fitzroya cupressoides* in southern Chile; new millennial, or longer, chronologies from Canada, the United States, Mexico, Chile, and Argentina; and a 700-year precipitation-sensitive record from *Polylepis tarapacana* in the Bolivian Altiplano. New temperature and precipitation reconstructions were developed throughout the transect. The more notable include 300-year temperature reconstructions from northern and southern Patagonia, a millennial summer temperature record from the Canadian Rockies, an 1,800-year El Niño–sensitive record from *Austrocedrus chilensis* in Argentina, and a 600-year precipitation reconstruction from Mexico. Drought and precipitation reconstructions in Canada, the United States, Mexico, Chile, Argentina, and potentially Bolivia, plus new stream-flow reconstructions from Canada and the first from Chile and Mexico, provide important long-term perspectives on the availability of water resources. Investigations in Canada, Chile, and Argentina have expanded the applications of tree rings to the study of past glacier fluctuations, including reconstructing changes in glacier mass balance. The "tropical gap" in high-elevation chronologies has been reduced to the area between 19°N and 16°S, and the first short lowland chronologies have been developed from Ecuador and Peru between 2°S and 7°S. Several new tropical species with annual rings have been identified in Mexico, Peru, Ecuador, and Bolivia that will allow the determination of tree age and will be useful in studies of forest dynamics and land use history (Roig et al. 2005, Rodriguez et al. 2005).

The establishment of chronology networks has allowed important reconstructions of El Niño, the Mexican Monsoon, and other atmospheric circulation patterns. The analysis of large-scale climate relationships shows strong linkages between temperature variability in coastal Alaska and Patagonia (mediated via the tropics), as well as significant relationships between reconstructed sea surface temperatures in the central Pacific over the last 300 years and precipitation/drought-related patterns in the southwestern United States and Mexico, the Bolivian Altiplano, and central Chile. These results show important changes in the behavior of the system prior to 1850 (Villalba et al. forthcoming). The results of this work have been published in several books or monographs and more than 130 referred journal articles and book chapters, and they have been delivered in several hundred conference and workshop presentations. At the outset of the project, we contributed to *Dendrocronología in América Latina* (Roig 2000), a review of previous work in South America that included the first systematic overviews of major techniques and applications of this science in Spanish.

Policy Relevance

Climate variability is a key parameter in global change, influencing all natural, human, and economic systems. Natural climatic patterns vary at all temporal and spatial scales, and future anthropogenic changes will be superimposed on, and greatly affected by, this natural variability. Long proxy climate data series with annual resolution are needed to assess the full range of climate variability. Tree-ring data and the climate reconstructions provided by them are the most ubiquitous and reliable source of such data. Understanding and documenting climate variability are key components of the major international projects such as Past Global Changes (PAGES), Climate Variability and Predictability (CLIVAR), and World Climate Research Programme (WCRP) as they provide policy-relevant information to decision makers for a better use of water and other natural resources. Our role here is seen in data provision and analysis using conventional academic outlets that target the global and regional scientific community where these data can be integrated into, for example, studies by the Intergovernmental Panel on Climate Change (IPCC), World Meteorological Organization (WMO), or national and regional agencies. The payoff is long term, not immediate, and thus difficult to estimate in terms of short-term impacts on the user community that may ultimately use these results over the next decades.

We have also undertaken applied studies involving forestry in several regions that relate to fire history, ecosystem conservation, management of natural reserves, and biomass yields in semiarid scrublands. Innovative studies have developed a growing body of work establishing linkages between climate, disease, famine, crop yields, and societal upheaval in Mexico and 19th-century Argentina. These results have policy implications at the national, regional, and local levels, where results and or implications of our studies were communicated to and applied by government agencies, nongovernmental organizations (NGOs), private companies, or research scientists. This has largely been carried out by the project's Co-PIs in their respective countries, developing a growing network of contacts.

Bridging the Gap between Natural and Social Sciences

As the research network was conceived as a "natural science–based" project, there was no explicit attempt to include social scientists as co-investigators, although there was recognition from the start of the need to develop an "applied" side to the activities of the project. Over one-third of the Co-PIs received graduate training or teach in geography departments, and therefore there was familiarity with, and some expertise in, social science disciplines. Moreover, paleoenvironmental sciences tend to be holistic and interdisciplinary rather than reductionist and experimental ("Class B sciences," according to Harvard University, Gould 1989); they often deal with case studies and unique situations, as commonly found in the historical Earth and life sciences. The ecological, hydrological, and forest sciences have applied aspects that involve integration and interaction with social scientists and research managers more than other science disciplines. Therefore, although the group has common language and concepts through dendrochronology and climate change research, they also brought other backgrounds to discussions, easing interdisciplinary communication.

Many natural and social scientists have stereotypical views of the other's science. A major problem is the refusal to enter collaboration with an open mind. Disciplines closet themselves in a coded language (technical terms, equations, concepts) that baffles outsiders. Key words and concepts have different meanings for different disciplines (this is doubly difficult in a second language, where "normal" words carry disciplinary meanings that are not found in a standard dictionary). Interdisciplinary research must tear down these barriers and work in (or at least accept) a common language. Disciplines are taught to address their questions in certain ways and have different standards, protocols, and ways of carrying out research. The most extreme ends of the continuum are the differences between natural and social sciences, but such differences also occur within a science and even within disciplines. The key is to accept, evaluate, and respect all approaches to a problem and to eschew disciplinary hegemony in the interest of communication.

A co-investigator on this project provided an example of this problem: I chose a term because it is specific to my discipline and those in my discipline understand it. Maybe it wasn't a good idea because you (i.e., those in other disciplines) did not have this background. Perhaps it is better to work with data (and terms) that everybody can understand, but it would not be as useful (or acceptable) to my colleagues.

In a similar vein, when asked to comment on the implications of a historical study, the co-investigator said: We (historians) never speak about implications of this for future management. My teachers always underlined that I should not speak about what to do, but about what had happened, and I prefer to speak about what I know.

These two remarks illustrate the pressures that young researchers face in trying to break away from disciplinary mores and communicate research findings to a broader interdisciplinary audience, other researchers, and even the general public.

These difficulties are compounded by the need for young scientists to publish in "journals that count" (ISI journals) that often are discipline based. Many disciplines have their own particular sets of rules about what are acceptable publications (and journals) and are somewhat intolerant of other approaches. This "my way or the highway" syndrome is often counterproductive. Papers that attempt to meld social and physical science approaches may encounter these problems at the publication stage and be rejected by reviewers from "journals that count" in the authors' fields because they diverge from these accepted modes. As these first publications are often essential building blocks in academic careers, young scholars may chose to place papers in the journals familiar to their disciplinary peers rather than coauthor papers outside that field. Although working with colleagues of different backgrounds can (and does) produce exciting synergy, it may also be frustrating when disciplinary barriers are raised. Destructive criticism hurled from one ivory tower to another does not help build bridges between them. There is a need for mutual respect and trust. Reactions to other disciplines should be based on observations of what these individuals contribute and not on stereotypes derived from other sources that place artificial boundaries around subject areas.

Outreach from the Project: Who Were Our Stakeholders and How Did We Involve Them?

In this project, we have recognized three groups of stakeholders—scientists, practitioners, and students. The major stakeholders have been the scientific community of paleoenvironmental, climate change, and other global change scientists who address the issues of climate change, climate modeling, and impacts. We have provided data, reconstructions, and other materials that can be integrated in local, regional, and global studies and also in modeling. They have been communicated via normal scientific channels—through papers, conference presentations, and personal contact and by archiving data in databases that are freely (and easily) accessible to scientists around the world, such as the National Oceanic & Atmospheric Administration (NOAA) paleoenvironmental databases. For example, some of the more recent attempts to reconstruct Northern Hemisphere temperature records of the last millennium include data that we have developed. Similarly data from CRN studies have been incorporated in a new, gridded, drought severity data set for Canada, the United States, and Mexico that lays the foundation for many future significant studies of continental-scale droughts (Cook et al. 2004). Such data are also highly significant at regional and local levels for reconstructions of precipitation and drought frequency and, through stream flow reconstructions, provide valuable data for long-term water management planning. Presenting these data to water managers and engineers at workshops in Mexico and the United States has created demand for these data for use in planning of water resource management and the assessment of future risk. Long-term studies of hydroclimatic variability are essential in developing sound water management strategies for a huge megalopolis like Mexico City, where 20 million people demand a water supply of over 65 cubic meters per second.

At a second level, we self-identified projects at the local or regional level that demonstrated applications of tree rings to address specific problems, met significant data needs, and provided insight and direction toward future policy and practices in several areas (Luckman 2005). These initiatives were undertaken by individual PIs or small research teams. Results have been communicated to the user sector through reports, scientific meetings, and workshops with local foresters, resource managers, government scientists, farmers, and others. Studies of semiarid *Prosopis* woodlands in the Monte of Argentina provided data on growth rates, wood production, and sustainable yields that are critical for the future management of this resource for charcoal, timber, and other forms of production (including, potentially, carbon sequestration). Similar studies of wood productivity in the mesquite woodlands of the Chihuahuan Desert (Mexico) provided essential data for the sustainable management of this ecosystem; the resulting publication (Villanueva-Diaz et al. 2004) was widely distributed to agencies dealing with the management and conservation of woodlands. The CRN participated in a workshop on the role of fire in forest ecosystems in Chile (funded primarily by CONAF, the Chilean forest service), involving university researchers and foresters who provided a scientific basis for the design of policy and action plans concerning forest fires in native forests. Also, at the national level, Co-PIs in Chile hosted a "national scientific meeting" on native forests in July 2003, involving an interdisciplinary group of university scientists and forestry professionals. Conclusions and recommendations from the meeting were presented to the Agriculture Commission of the Chilean senate and to the general secretary of the president of Chile. These were later published in a widely distributed book (Lara et al. 2003) that contains key recommendations for forest policy and forest legislation in Chile. The authors anticipate that these recommendations should have a significant impact on decision making regarding the native forests in Chile.

The climate-related work has also provided case studies of significant impacts of climate variability on human populations in the past, particularly with respect to drought, famines, disease outbreaks, and changing crop yields in Mexico (Acuña-Soto et al. 2002; Therrell et al. 2006). Similarly, investigations of forest tree populations have provided significant inputs to the management of natural reserves and been used by NGOs to promote the conservation of fragile ecosystems.

Through individual laboratories, the research network has provided training in dendrochronology to other networks. When Hurricane Isidore in 2002 felled innumerable trees in Yucatán, Mexico, CRN researchers working in that region initiated a joint salvage dendrochronology project. This resulted in a major reference monograph, currently in press at Universidad Autónoma Yucatán, describing the characteristics of many species new to dendrochronology, some of which will prove useful in studies of forest dynamics and land use history simply by providing the capability to age the trees in stands or garden plots. Similar studies, for example in Bolivia and Peru, provide data on other "new" tropical species that have similar potential.

The key to success identified here is to work locally and regionally—where the individual or group can make contacts and lasting collaborative relationships with policy makers and managers that will survive long after the initial flush of funding. Workshops, open houses, and small meetings demonstrate the utility of the techniques in the local setting. Local knowledge of the way the system works is needed. Having some Co-PIs in significant positions in government agencies may also help provide entry into adoption and application of this work and also help shape the institutional demand for the research.

Capacity Building

The final stakeholder group we identified was made up of students and junior researchers. An underlying goal of our project's activities was to promote our science and its utility in global change studies. In addition to publicity and dissemination of results, it is important to recruit and train new individuals in the field. The costs involved in training doctoral students outside their host country (particularly in North America) are prohibitive on a limited budget and also high risk (with no guarantee of success or return to the home country). Therefore we focused on training *in situ*, using short (5- to 10-day) courses that involved 20 to 30 students acquiring the basic tools of dendrochronology or being exposed to studies of climate change in their region. These students were drawn from many institutions and disciplines, and if we are successful in persuading even 10 to 20 percent to continue with their studies, it is money well spent. We also funded short research/training visits abroad to develop contacts and networking by junior scientists, as well as salary/scholarship support for student participation in active research projects with experienced Co-PIs. This, in our experience, is the best way to attract bright young students into research in all fields. These individuals will move into teaching, academic positions, government agencies, NGOs, and even private industry with a set of skills and tools necessary to address some critical global change issues. They will be able to provide data that will lead to better decisions on water, forest management, and land use change. During the CRN project, we funded participation of over 105 students and 25 senior researchers from 11 IAI member countries in four training field weeks, and an equivalent or greater number attended short lecture or field courses in Chile, Bolivia, and Argentina. We produced a 20-minute video of one of these field weeks, dubbed in English and Spanish, which has been excellent promotional material for individual research programs and for the IAI, and which also has been used by the National Science Foundation. Our Mexican colleagues produced a short video on tree rings for primary school children.

A primary goal of the project was to expand the availability and use of facilities in Latin America. New laboratories were established in Durango, Mexico, and La Paz, Bolivia, in October 2000 and, using additional funds from the IAI, at Piura, Peru, in January 2001. These were the first tree-ring labs in Peru and Bolivia, and the first dendroclimate laboratory in Mexico. The Durango laboratory has recently been recognized as the National Dendrochronological Facility in Mexico. Dedicated IAI-funded staff were associated with each facility and have undergone training at one of the senior laboratories participating in the CRN. These laboratories have developed collaborative projects with scientists in their host countries and have organized workshops, training sessions, and education and outreach activities. In 2004 using a similar model, a fourth lab was established in Porto Alegre, Brazil. During the project, the Mendoza lab in Argentina has expanded and consolidated its role as the major tree-ring laboratory in the Southern Hemisphere. It has provided considerable training for Latin American scientists, both in house and by short courses delivered at other facilities in Argentina and elsewhere. The other laboratories involved in the project from Chile and Canada have enhanced their national profiles, international contacts, and research activities through collaboration within the CRN.

Additional Comments

Measuring the impact of our research is difficult: Simple conventional measures of short-term output cannot measure the longer-term impacts of, for example, data and database development, student training, and educating public awareness at all levels. It would be unrealistic, for example, to measure the impact of a research network by pages of peer-reviewed papers per Co-PI or per research dollar. For useful discussion of this and related issues, see Masseli et al. (2004). The results of our activity will continue to yield the scientific information and manpower that supports new research and will continue to provide policy-relevant information to decision makers. We realize that the scientist's job does not end when one simply puts the information "out there" for others to use, but there are strong limitations on individuals' time commitments. From a career scientist's viewpoint, despite lip service to the contrary, teaching and communication garner few academic kudos. But we do have important material to communicate and need to learn or have access to ways to publicize and gain support for this work. Every large research project possibly needs a publicist who can take acquired knowledge and put it into the public eye—some researchers are good at this, and others need assistance. One other critical function is management of the project itself. In our network, almost all resources went into science, networking, and training with little attention given to management costs and effort. However, managing the accounts of a multiyear research project through multiple institutions in five or six countries and currencies, with different financial climates and rigid reporting guidelines, is not a task to be undertaken lightly and may need dedicated support.

Challenges for the Future

Policy relevance has been enhanced through the networking both in outreach and capacity building, but this comes at considerate transaction cost. The IAI CRN experiment has expanded the scientific horizons, the network of contacts, and the research foci for the participants at all levels, from Co-PIs to undergraduate research assistants across some 13 IAI countries. Collaborative research across disciplines, institutions, and countries provides perspectives that lead to greater understanding and appreciation of differences in the approach to, and circumstances of, scientific research in different environments. It has allowed researchers access to a range of expertise that would not otherwise be available. To quote from one researcher, "My participation on this IAI project has been unparalleled and has provided an excellent opportunity to interact with highly qualified people working in different countries but focused in common issues." The IAI CRN has provided opportunities to students and younger scientists for exchange, training, and personal academic development that would not otherwise have been available and has spawned a generation of new scientists across Latin America. One key advantage noted by several Co-PIs is that the provision of secure funding for an extended period of time has allowed development of more complete and ambitious research programs. Some of the research teams involved had never previously had external funding, and other groups had never engaged in projects with a window of more than two years.

The evolution of our CRN at times has been frustrating, with delays and the continuing evolution of administrative frameworks as the IAI's international effort grew. However ,this is only to be expected in a new venture of this magnitude, and the benefits considerably outweigh these minor irritants. The IAI CRN initiative established a new infrastructure that has produced innovative, interdisciplinary, international science that has led to significant developments in paleoenvironmental and global change science in the Americas. All Co-PIs would agree that the continuation of this initiative is crucial to effective progress in the study of global change in the Americas and would welcome participation in future endeavors. This is both one of the great strengths and one of the potential problems that now face the IAI. Having built successful teams and projects, how can the IAI maintain this momentum while simultaneously developing new projects with limited funding opportunities? How can (or should) it manage the transition of its fledglings out of the IAI nest to become successful, mature, independent "adults" in the real (and vastly different) world of scarce national and international research funds in order to make space for the next generation of projects? It might be useful to maintain a longitudinal study that monitors these outcomes over the next five years, to assess the impact of the CRN experiment on global change science in the Americas.

Literature Cited

Acuña-Soto, R., D. W. Stahle, M. K. Cleaveland, and M. D. Therrell. 2002. Megadrought and megadeath in 16th century Mexico. *Emerging Infectious Diseases* 8(4):36–2.

Cook, E. R., C. Woodhouse, C. M. Eakin, D. M. Meko, and D. W. Stahle. 2004. Long-term aridity changes in the western United States. *Science* 306:1015–8.

Gould, S. J. 1989. The Burgess Shale and the nature of history. In Gould, S. J., ed. *Wonderful life*. New York: W.W. Norton. (pp. 277–82)

Lara, A., D. Soto, J. Armesto, P. Donoso, C. Wernli, L. Nahuelhual, and F. Squeo, eds. 2003. *Key scientific components for a national policy on the uses, services and conservation of the Chilean native forests.* Valdivia: Universidad Austral de Chile, Iniciativa Científica Milenio de Mideplan. (134 pp.)

Luckman, B. H., ed. 2005. Application of dendrochronology to human dimensions related global change issues. Special Issue. *Dendrochronologia* 22(3):131–240.

Maselli, D., J.-A. Lys, and J. Schmid. 2004. *Improving the impacts of research partnerships.* Berne: Swiss Commission for Research Partnerships with Developing Countries (KFPE). (85 pp.)

Rodríguez, R., A. Mabres, B. H. Luckman, M. Evans, M. Masiokas, and M. K. Ektvedt. 2005. "El Niño" events recorded in dry-forest species of the lowlands of northwest Peru. *Dendrochronologia,* 22:181–6.

Roig, F. A., ed. 2000. *Dendrocronología in América Latina.* Mendoza: Universidad Nácional de Cuyo (EDIUNC). (431 pp.)

Roig, F. A., J. J. Jiménez Osornio, J. Villanueva-Diaz, B. H. Luckman, H. Tiessen, A. Medina, and E. J. Noellemeyer. 2005. Anatomy of growth rings in the Yucatán Peninsula. *Dendrochronologia* 22:187–93.

Therrell, M. D., D. W. Stahle, J. Villanueva-Diaz, E. Cornejo-Oviedo, and M. K. Cleaveland. 2006. Maize yield in central Mexico: 1474–2002. *Climatic Change* 74:493–504.

Villalba, R., B. H. Luckman, J. A. Boninsegna, R. D. D'Arrigo, A. Lara, J. Villanueva-Diaz, M. Masiokas, J. Argollo, C. Soliz, C. LeQuesne, D. Stahle, F. Roig, J. C. Aravena, G. Wiles, and P. Hartsough. Forthcoming. Dendroclimatology from regional to continental scales: Understanding local processes to reconstruct large-scale climatic variations across the western Americas. In Hughes, M. K., T. W. Swetnam, and H. F. Diaz, eds. *Dendroclimatology: Progress and Prospects.* Springer.

Villanueva-Diaz, J., R. Jasso-Ibara, G. Gonzáles-Cervantes, I. Sánchez-Cohen, and M. C. C. Potisek-Talavera. 2004. El Mezquite en la Comarca Lagunera: alternativa de producción integral para ecosistemas semidesérticos. *Folleto Científico No 14.* Durango, Mexico: INIFAP CENID-RASPA, Gomez Palacio. (35 pp.)

13

Climate Variability and Its Impacts in Mexico, Central America, and the Caribbean

Victor Magaña and Ernesto Caetano

The main objective of our collaborative research network (CRN) was to improve understanding of regional climate variability in Mexico, Central America, and the Caribbean in order to provide more accurate and adequate climate predictions to important socioeconomic sectors. From the climatic point of view, the region of interest is unique because of its complex topography and its two surrounding oceanic warm pools, one in the tropical Northeast Pacific and one in the Intra-Americas Sea. From the socioeconomic point of view, the various countries in the region share a common cultural background and exhibit similar levels of vulnerability to climate conditions. Therefore, an important element of the study was that the results would be used toward regional adaptation measures.

Activities in this project began in June 2000 with a field experiment in which meteorologists, biologists, chemists, oceanographers, and other experts in climate science interacted. Sharing information obtained during cruises in the warm pools of the Intra-Americas Sea led to a multidisciplinary description of the region's climate variability and its impacts on various socioeconomic sectors. Most work focused on the coordination and development of the Climate Experiment in the Warm Pools of the Americas (ECAC). The experiment was based on hypotheses about the onset of the rainy season and the occurrence of the midsummer drought (MSD) (Magaña et al. 1999). Activities included the following:

- empirical studies to better understand the climate dynamics of the region
- capacity building in the use of numerical models for hypothesis testing
- training and education of students through participation in the field experiment and interpretation of results
- multidisciplinary interactions to place climate information in a socioeconomic context

In order to test our hypotheses on the causes for the MSD, we monitored atmospheric and oceanographic factors considered crucial to the summer rains by collecting and analyzing four sets of data, three in the Northeast Pacific and one in the Caribbean. The results established that the warm pools of the Intra-Americas Sea are key elements that regulate characteristics of the summer rains. The main results of these analyses have been published (Ambrizzi and Magaña 2002; Caetano et al. 2002; Magaña et al. 2002) and can be summarized as follows:

- The meridional gradient in sea surface temperature (SST) over the eastern Pacific is an important factor in predicting the onset of the rainy season over Mesoamerica (Uribe 2002).
- The MSD is influenced not only by local SST variations but also by the circulation anomalies that result in SST fluctuations (Magaña and Caetano 2005).
- Current general circulation models cannot precisely reproduce the MSD, because of their coarse resolution, limiting the capacity to seasonally predict rains. Regional climate models may serve to produce regional climate predictions (Magaña and Quitanar 1997).
- There is enough seasonal predictability in the region to make systematic climate predictions possible and beneficial (Uribe 2002; Magaña 2005).

As part of the Climate Outlook Fora in Central America and Mexico, better ways of transmitting climate information have been explored. Clearer information is necessary to implement warning systems and mechanisms aimed at ameliorating the negative impacts of extreme climate conditions. Community Climate Model 3 (CCM3) and the Mesoscale Model (MM5) are examples of climate information based on numerical models that are being prepared at the regional level for the disaster preparedness and the agricultural sectors.

Outreach activities included capacity building in various socioeconomic sectors, such as the insurance sector in Mexico. A one-week training course for more than 20 professionals working in the insurance and agricultural sectors explored probabilistic seasonal climate prediction, uncertainty, and decision making.

As part of a major capacity-building effort, researchers and students from the CRN are taking part in the Megacities Impacts on Regional and Global Environments–Mexico City (MIRAGE-Mex) case study, which involves coordinated aircraft- and ground-based measurements supported by extensive modeling and satellite observations. This experiment is an intensive observational mission to study the chemical and physical transformations of gaseous and aerosol pollutants in the outflow of the world's second-largest metropolitan area, Mexico City. Significant opportunities for education and capacity building are expected.

One of the most important achievements of the CRN, in terms of interactions with social scientists, has been the capacity to implement joint activities aimed at communicating climate information for the region and evaluating the impacts of climate change. Frequent workshops with anthropologists and economists have addressed the value of seasonal climate predictions. Anthropologists have examined from a historical perspective the signals related to El Niño in disasters affecting Mesoamerica. This cooperative work has involved investigators from two CRNs that share the same interests.

The existence of collaborative research between natural and social scientists has brought together a group of scientists who also respond to requests for information from the Mexican administration. As a result, the Mexican authorities have allocated resources to problems posed by climate change. Since Mexico is in favor of initiatives such as the Kyoto Protocol, there is great demand from the Mexican government for knowledge about the implications of pronouncements by the international Climate Change Fora. The interaction of the Ministry of the Environment and our CRN has resulted, after some years of collaboration, in an initiative to create the Mexican Program on Climate Modeling. This project has resulted in a science plan, which is currently evolving into an implementation plan.

This initiative includes interaction with international groups toward the design of an atlas of the climate of the region. Prepared with the University of the West Indies (Jamaica), the atlas will update information on precipitation, temperature, and winds used as reference data for climate change studies. With collaboration from the University of Costa Rica, numerical models for regional seasonal weather predictions are being implemented. This joint effort has built capacity in Mexico and Costa Rica to prepare climate information for the region. Models are currently being evaluated for their predictive performance.

The Mexican and US governments have implemented a project on adaptation to climate change in the urban water sector of Hermosillo, Sonora, Mexico. Thanks to the experience in collaborative work within the CRN, the project was conducted with the involvement of consultants from the USA and anthropologists working at the University of Mexico. In close interaction with stakeholders of the water sector in Hermosillo, a feasibility study was conducted to see which adaptation options could be implemented to improve problems in the urban water sector. Proposed adaptation measures included an educational and media program on the culture of water, means for enhanced infiltration that remain effective even during extreme precipitation events, and a building design that uses new materials to cool houses. The study results have been presented to the new local authorities for consideration and adaptation (Eakin et al. 2005).

The interaction with stakeholders, particularly local government, was established in 1997. At that time, a major El Niño event had just occurred and there was great interest in understanding the causes of a major drought throughout most of central Mexico. Nongovernment organizations, farming associations, and environmentalists from Tlaxcala have an interest in seasonal climate predictions at a regional level that could help them prepare to face another El Niño situation. The prediction for the summer of 1998 was accurate enough for stakeholders to become more interested in using climate information. Since then Tlaxcala has become a sort of laboratory, not only working with climate prediction but also exploring methods of decision making in sectors such as agriculture and forestry.

The CRN's experience with water stakeholders led the Mexican authorities to involve the CRN's scientists in a study on capacity building for adaptation to climate change. This project involves similar developments in Central America and Cuba. The experience acquired within this CRN has been fundamental to developing close collaborations with other scientists in the region and to sharing knowledge, data, methodologies, and results that will be of use to the governments of the region in establishing national climate change programs. Results from our study will serve to construct the Mexican national climate change program and action plan. That is, they will serve to design national policies on climate change issues.

An important element of this science project was the interaction with decision makers. As a result, capacity was generated not only among students but also among government administrators. The scientific information generated by the fields of climate and social sciences was disseminated through conferences, workshops, meetings, and publications and also popularized in books and other publications. A constant effort to inform the public about the CRN's research on climate variability and change resulted in increased awareness, for instance, in the agriculture and forestry sectors.

We can conclude that, before the establishment of this CRN, most scientists involved in global change studies focused narrowly and collaborated mainly with colleagues from their own countries or the United States and Europe. As a result of the CRNs of the Inter-American Institute for Global Change Research (IAI), closer collaborations were established between Latin American scientists and many new joint studies were conducted. In addition, and perhaps more importantly, researchers in various fields came together to develop multidisciplinary work aimed at solving problems of relevance in the environmental sciences. Maintaining communication with their national authorities guided scientists toward addressing scientific problems relevant to the decision-making processes in national and international fora. Some of the most important elements of the IAI CRN projects involved capacity building, preparing students at all levels, and encouraging scientists to interact with stakeholders and decision makers. At present some of those students work in various sectors in which climate information or other types of environmental information are necessary for making decisions. Capacity building has been proven a sound strategy for educating governments on the importance of science in their sector.

Literature Cited

Ambrizzi, T., and V. Magaña. 2002. Experimento Climático Nas Aguas Quentes da America Central. *Boletim da Sociedade Brasileira de Meteorologia (SBMET)* 22:6–12. (in Portuguese)

Caetano, E., V. Magaña, and J. L. Vásquez. 2002. Observational study of the middle summer drought during the climate experiment in the Americas warm pools. *25th Conference on Hurricanes and Tropical Meteorology, San Diego, California.* Boston: American Meteorological Society. (pp. 559–60)

Eakin, H., V. Magaña, J. Smith, J. L. Moreno, J. M. Martínez, and O. Landavazo. 2005. A stakeholder-driven process to reduce vulnerability to climate change in Hermosillo, Sonora, Mexico. Forthcoming in *Mitigation and Adaptation Strategies for Global Change.* Springer.

Magaña, V. 2005. Programa de Modelación del Clima en México. Mexico City: Mexico National Institute of Ecology.

Magaña, V., J. A. Amador, and S. Medina. 1999. The mid-summer drought over Mexico and Central America. *Journal of Climate* 12:1577–88.

Magaña, V., and E. Caetano. 2005. Temporal evolution of summer convective activity over the Americas warm pools. *Geophysical Research Letters* 32:L02803, 10.1029/2004GL021033. (January 19, 2005)

Magaña, V., E. Caetano, and J. L. Vásquez. 2002. Climate experiment in the Americas warm pools. *25th Conference on Hurricanes and Tropical Meteorology, San Diego, California.* Boston: American Meteorological Society. (pp. 517–8)

Magaña, V., and A. Quintanar. 1997. On the use of a general circulation model to study regional climate. In *Proceedings of the Second UNAM-CRAY Supercomputing Conference on Earth Sciences, Mexico City.* New York: Cambridge University Press. (pp. 39–48)

Uribe, E. M. A. 2002. El inicio de la Temporada de Lluvias en la Costa Sudoeste de México: Relaciones para su Diagnóstico y Pronóstico. MSc dissertation, National Autonomous University of Mexico. (69 pp., in Spanish)

14

Stakeholders and Decision Makers in a Study of Global Changes in the South Atlantic

Edmo J. D. Campos and Alberto R. Piola

The Science Questions

The ocean plays a key role in the climate system by redistributing excess heat from low to high latitudes and by storing large amounts of heat and greenhouse gases. Within the South Atlantic, shallow and deep flows of the global thermohaline circulation are modified by intense mixing and by sea-air exchanges. Nutrient inputs from continental outflows and internal recycling support substantial primary production in the coastal ocean, which therefore exchanges much larger amounts of carbon dioxide (CO_2) with the atmosphere per unit area than the deep ocean and also supports significant fisheries (Piola et al. 2005). In recent decades, rates of coastal urbanization, industrial development, and other human activities have increased considerably, altering the delicate physical-biological balance of coastal oceans worldwide.

Recent reports suggest that global warming may be the primary cause of the collapse of coastal environments. Therefore a better understanding of the dynamics of coastal and deep oceanic regimes, and how these environments may influence and are affected by global changes, is an enormous scientific challenge, requiring coordinated multidisciplinary efforts, including ocean and atmospheric physics and chemistry as well as marine biology.

The water masses of coastal oceans containing organisms and dissolved chemicals extend across political boundaries. The scientific understanding of these highly variable cross-boundary regimes requires a high level of international coordination. That is to say, regional approaches to the study of the coastal ocean are not possible without the establishment of solid scientific ties and strong commitment from governments. The

Box 14.1. Key scientific findings

The scientific research conducted by SACC scientists, based on observational work and analysis of historical oceanographic data, atmospheric reanalysis, remote sensing, and numerical models, has contributed to advancing the knowledge of the South Atlantic and its role in environmental and climatic conditions in South America. Some of the most important of these findings can be summarized as follows:

- synoptic documentation, for the first time ever, of the seasonal variations of the Plata Plume and the subsurface frontal structure between subantarctic and subtropical waters over the shelf (Piola et al. 2000; Piola et al. submitted; Möller et al. submitted)
- confirmation that the frontal structure has direct impact on the distribution of plankton and icthyoplankton, illustrating how physical characteristics impact the distribution of biota, including commercially important species, producing baseline data necessary for resource management
- confirmation that fluctuations of continental precipitation, river outflow into the ocean, and surface winds control the distribution of coastal waters, which is correlated to sea surface temperature (SST) variability (Campos et al. 1999; Piola et al. 2000; Piola et al. 2005; Piola et al. submitted)
- basin scale modeling and analysis of existing observations, helping to identifying the mechanisms leading to the formation of the dipole SST structure that dominates the South Atlantic at interannual to decadal timescales; also revealing that the SST variability is strongly correlated with the Atlantic Inter-Tropical Convergence Zone (ITCZ) and with anomalies of the South Atlantic Convergence Zone (SACZ) in the atmosphere, which in turn are linked to variations in the precipitation regime over central South America (Palastanga et al. 2002; Haarsma et al. 2003; Haarsma et al. 2005; Chaves and Nobre 2004)
- enhancement of the forecast skill of the SACZ, relative to atmospheric general circulation model (GCM) hindcast experiments forced by observed SST fields, through our coupled ocean-atmospheric modeling research (Chaves and Nobre 2004)
- confirmation that, excluding the upwelling regimes along the western African coast, the Patagonian shelf break is the region with the largest primary production in the South Atlantic (Saraceno et al. 2005; Romero et al. 2006) and where large amounts of CO_2 are absorbed by the ocean (Bianchi et al. 2005)

Our efforts have contributed to the design of a South Atlantic observation system for the monitoring of long-term variability.

South Atlantic Climate Change Consortium (SACC) is a successful example of such synergistic international cooperation.

Created in 1996 with the support of an Initial Science Program grant provided by the Inter-American Institute for Global Change Research (IAI), the SACC began as a small cooperative of South and North American scientists with common interest in the South Atlantic. In 1999, with the help of a Collaborative Research Network (CRN) grant, it evolved into a strong research consortium. SACC's scientific objectives included understanding the impacts of climate change on western South Atlantic coastal ecosystems and the role of shelf sea surface temperature (SST) variations on the regional climate. To accomplish its goals, SACC has fostered the international cooperation that facilitated observations over the continental shelves of Argentina, Brazil, and Uruguay.

IAI and other agencies from the participating countries provided financial support for the cooperative network of oceanographers, meteorologists, and climatologists and its research activities. These activities included the development of an observational and numerical modeling effort without precedent in the region and provided significant capacity building. Lately the network has been extended to include scientists from institutions in Chile and Peru. Key scientific findings of the SACC are listed in Box 14.1.

Outreach, Networking

SACC's findings were communicated to the general public in media interviews, newspaper articles, science magazines, and newsletters. The main aim of the outreach activities of this science program, though, was the establishment of international consortia that combine national and international concerns and interests in oceanography and ocean monitoring, because the oceans transcend national boundaries, jurisdictions, and concerns.

Variability of the equatorial Atlantic SST influences the precipitation regimes over low-latitude South America and Africa. The Pilot Research Moored Array in the Tropical Atlantic (PIRATA) cooperative project between Brazil, the United States, and France was launched in 1997 to monitor ocean variability at selected tropical locations. Soon after its implementation, it was recognized that extratropical SST variability in the western South Atlantic was associated with variations in the South Atlantic Convergence Zone. Recognizing a synergism with PIRATA, SACC promoted and cosponsored the first PIRATA Western Extension (WE) Workshop. This meeting gathered over 40 scientists from the Americas, Africa, and Europe to discuss the extensions of the PIRATA array. SACC investigators prepared the scientific rationale for a western extension, which was approved by the project's International Scientific Steering Group in 2004 and funded by the Brazilian Ministry of Science and Technology in 2005.

A workshop on South Atlantic climate observing systems, held to discuss advances

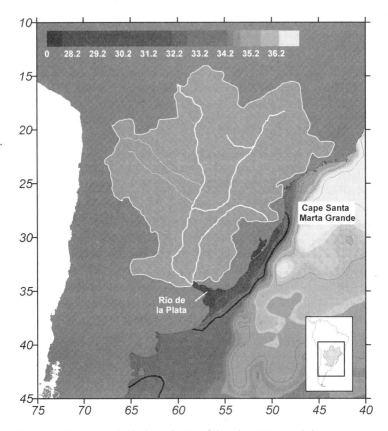

Figure 14.1. The hydrologic basin of the Plata River and the western South Atlantic

in the knowledge of the role of the South Atlantic on climate, provided a unique opportunity to integrate natural scientists with regional stakeholders and decision makers. The meeting of more than 100 scientists from the Americas, Europe, and western Africa addressed the South Atlantic's influence on the meridional overturning circulation and on global and regional climates. It produced a comprehensive set of recommendations resulting in the following actions:

• Brazil, Argentina, and South Africa joined the Argo program, a global array of 3,000 free-drifting floats that monitor the temperature and salinity of the upper 2,000 m of the ocean, with all data being relayed and made publicly available within hours after collection. In May 2005 the South Atlantic Argo Data Assemblage Center was created with the active participation of SACC investigators. In 2005 the first Brazilian and Argentine Argo floats were deployed by SACC investigators in cooperation with the National Oceanic and Atmospheric Administration (NOAA) and the Argentine and Brazilian navies.

- The Plata Plume Project (http://glaucus.fcien.edu.uy/pcmya/sacc/LaPlataS2004/index.html) promoted by the consortium made it possible, for the first time ever, to provide quasi-synoptic observations and modeling along the continental shelves of the three neighboring countries, Argentina, Uruguay, and Brazil, from 40°S to 25°S (Figure 14.1). This successful cooperation included the Argentine, Brazilian, and Uruguayan navies, the Uruguayan Air Force, and the Brazilian Ministry of Foreign Relations. In 2004 the navies and scientific institutions of Argentina, Brazil, and Uruguay signed a letter of intent for the creation of a Southwest Atlantic Regional Alliance related to the Global Ocean Observing System (GOOS). The Plata Plume Project involved a large number of scientists and students from institutions from all SACC participating countries (Argentina, Brazil, Uruguay, and the United States), including marine biology, chemistry, and physics experts.

Capacity Building

SACC organized training activities for young scientists in fieldwork techniques, data analysis, and numerical modeling, including international short courses; undergraduate, graduate, and postdoctoral fellowships; and field training. The short courses provided advanced training on specific subjects that were understudied in Latin America. The courses fostered the exchange of over 80 students from IAI member countries. Such contacts are essential for maintaining a global change research network in the region.

Challenges for the Future

Our analysis of *in-situ* and remote sensing observations has highlighted the functioning of a variety of frontal systems over the shelf and in the open ocean. The Brazil/Malvinas Confluence presents variations at seasonal time scales. These fluctuations are important because, as the confluence marks the southernmost extent of subtropical waters in the Southern Hemisphere, it is a significant source for a variety of subtropical mode waters, which fill the upper portion of the ventilated thermocline in the subtropical gyres of the Southern Hemisphere. Subduction of mode water carries surface properties, including CO_2 and other greenhouse gases, to the subsurface ocean. In addition, meridional shifts of the confluence cause large sea-air heat flux variations, which are believed to modulate regional precipitation anomalies over the productive agricultural plains of southeastern South America. Our analysis has improved our understanding of the regional climate variability and the oceanic variability associated

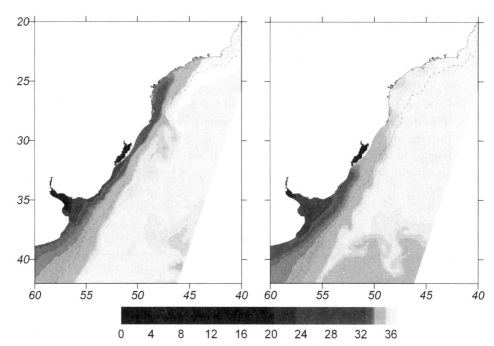

Figure 14.2. *Left:* Schematic circulation in the western South Atlantic shelf. *Right:* SeaWiFS January chl-a climatology depicting the biologically important regions, which coincide with the Magellan, Shelf Break, Rio de la Plata and Sub-Tropical Shelf ocean fronts, and the central Santa Marta and southern Brazil Bight.

with boundary ecosystems. Basin scale modeling and analysis revealed that the SST variability is linked to variations in the precipitation regime over central South America (Chaves and Nobre 2004). A follow-up CRN project aims to advance our understanding of the physical mechanisms that control biological productivity and biodiversity in highly productive regions of the western South Atlantic (Figure 14.2).

Literature Cited

Bianchi, A. A., L. Biannucci, A. R. Piola, D. Ruiz Pino, I. Schloss, A. Poisson, and C. F. Balestrini. 2005. Vertical stratification and air-sea CO_2 fluxes in the Patagonian shelf. *Journal of Geophysical Research* 110(C7):C07003 (10.1029/2004JC002488).

Campos, E. J. D., C. A. D. Lentini, J. L. Miller, and A. R. Piola. 1999. Interannual variability of the sea surface temperature in the South Brazil Bight. *Geophysical Research Letters* 26(14):2061–4.

Chaves, R. R., and P. Nobre. 2004: Interactions between the South Atlantic Ocean and the atmospheric circulation over South America. *Geophysical Research Letters* 31 (3):L03204 (10.1029/2003GL018647).

Haarsma, R. J., E. J. D. Campos, W. Hazeleger, C. Severijns, A. R. Piola, and F. Molteni. 2005. Dominant modes of variability in the South Atlantic Ocean: A study with a hierarchy of ocean-atmosphere models. *Journal of Climate* 18(11):1719–35.

Haarsma, R. J., E. J. D. Campos, and F. Molteni. 2003. Atmospheric response to South Atlantic SST dipole. *Geophysical Research Letters* 30(16):1864 (10.1029/2003GL017829).

Möller, Jr., O. O., A. R. Piola, A. C. Freitas, and E. J. D. Campos. Submitted. The effects of river discharge and seasonal winds on the shelf off southeastern South America. *Continental Shelf Research.*

Palastanga, V., C. S. Vera, and A. R. Piola. 2002. On the leading modes of sea surface temperature variability in the South Atlantic Ocean. *CLIVAR Exchanges*, no. 25, September 2002.

Piola, A. R., E. J. D. Campos, O. O. Möller, Jr., J. Charo, and C. Martinez. 2000. Subtropical Shelf Front off eastern South America. *Journal of Geophysical Research* 105(C3):6565–78.

Piola, A. R., R. P. Matano, E. D. Palma, O. O. Moller, and E. J. D. Campos. 2005. The influence of the Plata River discharge on the western South Atlantic shelf. *Geophysical Research Letters* 32(1):L01603 (10.1029/2004GL021638).

Piola, A. R., O. O. Moller, Jr., R. A. Guerrero, and E. J. D. Campos. Submitted. Seasonal variability of the Subtropical Shelf Front off eastern South America: Winter 2003 and summer 2004. *Continental Shelf Research.*

Romero, S. I., A. R. Piola, M. Charo, and C. A. E. Garcia. 2006. Chlorophyll-a variability off Patagonia based on SeaWiFS data. *Journal of Geophysical Research* 111(C5):C05021 (10.1029/2005JC003244,2006).

Saraceno, M., C. Provost, and A. R. Piola. 2005. On the relationship between satellite retrieved surface temperature fronts and chlorophyll-a in the western South Atlantic. *Journal of Geophysical Research* 110(C11):C11016 (10.1029/2004JC002736).

15

Climate Variability and Climate Changes in the Southern Cone

Mario N. Nuñez

Human activities are very sensitive to the timing and intensity of rain. This sensitivity is especially evident in much of the subtropics and tropics, where the majority of the annual rainfall occurs in clusters of heavy showers and thunderstorms. Relatively small changes in wind, moisture, temperature, and sea surface temperature fields can produce large changes in regional and local rainfall. Precipitation predictability on intra-seasonal and interannual time scales requires a thorough understanding of rainfall mechanisms. It is important to know how transient weather activity may be affected by climate variability and climate change.

Although many scientists within the Southern Cone focus on climate variability in the region and on the impact of that variability on the human population, for many years there has been inadequate collaboration and cooperation among these scientists. Yet the scientific problems are so complex that it is clearly beyond the capability of any single research group or institution to effectively address them. Among the problems that inhibit collaboration are the necessity for interaction between different disciplines (e.g., meteorology and social science), the historical resistance to the open exchange of data, and language barriers. The Inter-American Institute for Global Change Research (IAI) framework provided an opportunity to develop an integrated, interdisciplinary, and international project to address climate variability and climate changes in the Southern Cone.

The primary goal of this project was to advance the knowledge of regional climate variability and change, their prediction, and their impacts, focusing on 1) the role of large-scale sea surface temperature variations in determining climate variability in southeast South America, 2) tropical-extratropical interactions related to the circulation and precipitation variability over the area, 3) impacts of climate variability on sectors of social and economic importance in the Southern Cone, and 4) climate change scenarios for South America. The purpose of the project was to foster interaction by sponsoring regular meetings of the participants, as well as to promote long-term collaboration by virtue of a postdoctoral program, with the Collaborative Research Network sponsoring extended fellowships at the participating institutions. The emphasis of the postdoctoral program was on international exchanges. The project's 20 co-principal investigators came from 13 institutions in Argentina, Brazil, Paraguay, Uruguay, and the United States. This group included scientists from six universities, one government agency, and two research laboratories. Research was carried out on the floods in the La Plata Basin, on the physical processes of extreme events in the area, and on the degree of understanding of these extreme events by stakeholders (Caffera 2003). Regional studies of the low-level jet stream in South America, modeling of regional climate variability, and Atmospheric Global Circulation Models (AGCMs) were used to analyze the influence of the Pacific and Atlantic sea surface temperatures on the precipitation over the La Plata Basin. Climate change simulations of southern South America's climate were performed using a regional climate model nested in a time slice of long-term (240 years) Atmospheric General Circulation Model (AGCM) experiments.

Policy Relevance and Capacity Building

The MERCOSUR (Southern Common Market) region of South America is home to more than 200 million people. That population is profoundly affected by both short- and long-term climate variability. In the short term, floods and droughts result in deaths, displacement, and crop failures, with their economic and social consequences. Long-term variability can have impacts on agricultural production and its associated economies. In particular, systematic changes in the moisture and heat fluxes into the region, due either to anthropogenic effects or to natural climate variability, affect regional agriculture as well as the availability of water resources for energy production and other uses. Collaboration improved the capacity of the researchers to transfer usable knowledge, designed to address specific problems, to the benefit of the stakeholders.

Human actions are starting to interfere with global climate. Climate change and global warming are on the current political agenda. The growing interest in global climate change is due to the vital practical importance of this problem. The controversial assessment of anthropogenic contributions to climate change necessitates analysis of numerical modeling results and observational data on present climate changes and the causes of theses changes. This project's researchers provided climate change projections for the 2080 decade as a contribution to the Second National Communication of Argentina for Climate Change (SNCACC). Collaborators in Brazil and Uruguay have also performed numerical experiments to anticipate future climate change in their regions (Obregón and Nobre 2003).

This project helped improve the understanding of climate variability and prediction in the region, resulting in predictive capability and an understanding of how to best use those predictions. Training and education have been promoted through more than 30 fellowships funded by the IAI and regional funding agencies. Students were able to participate in exchanges between laboratories. Investigators from the United States, Argentina, and Brazil collaborated to produce a set of daily records from more than 2,000 stations south of 15°S with a uniform format. These data were obtained from various public and private sources in Brazil, Paraguay, Argentina, and Uruguay. While station density varies substantially in space, and the quality of the records differs, most records begin in the mid-1970s and continue until near the end of the last millennium. The data have already been useful in several diagnostic studies.

As a consequence of its results, the project has become an affiliated program of the Climate Variability and Predictability (CLIVAR) study, an interdisciplinary part of the Global Climate Research Program (GCRP) focusing on the variability and predictability of climate.

Literature Cited

Caffera, R. M. 2003. Primeros avances sobre eventos de inundación en Uruguay y la circulación atmosférica en capas bajas. (First approaches on flood events in Uruguay and low level atmospheric circulation.) *Meteorológica* 28(1 and 2):97–108.

Obregón, G. O., and C. A. Nobre. 2003. Rainfall trends in Brazil. *Bulletin of the American Meteorlogical Society* 84(8):1008–9.

16

Land Use Change in Semiarid Americas: Biogeochemistry, Societal Impact, and Policies

Rômulo S. C. Menezes, Elke Noellemeyer, Ignácio H. Salcedo, Juan J. Jiménez-Osornio, and Holm Tiessen

Many semiarid regions in the Americas sustain large populations that live at or below the poverty line. Food security and income generation for these populations depend on scarce and highly variable rainfall. Increasing pressure on the land has led to significant changes in land use and cover, followed by extensive degradation of natural resources and socioeconomic conditions. In other areas, due to off-farm income opportunities created by globalization processes, agricultural lands are being abandoned.

Knowledge about the drivers and decision processes involved in land use change and its impact on environmental quality and human livelihoods is a prerequisite for sustainable development in semiarid regions of the Americas. There is a need for developing mechanisms to ensure that scientific knowledge reaches societies and their political decision makers, raising awareness of the importance these changes have for the livelihoods of rural populations and for regional development.

In 1999 research institutions in Argentina, Brazil, Canada, Mexico, and Venezuela established the collaborative research network called Biogeochemical Cycles and Land Use Change in Semiarid Americas, with the main objectives of integrating biogeochemical and socioeconomic data at the farm and regional scale within different land use and soil management systems, in order to evaluate and predict regional changes in resource quality in the context of global economic and environmental changes.

Research activities were conducted in the semiarid pampas of Argentina, the semiarid northeast of Brazil, and the Yucatán peninsula of Mexico. A brief description of the overall environmental and socioeconomic characteristics of these study regions is presented in Table 16.1.

Table 16.1. Description of the study regions

Study Regions	Environmental Characteristics
La Pampa, Argentina	Area is approximately 140,000 km², with warm temperate, semiarid climate. Mean annual temperature is 16°C. Mean annual rainfall is 400–600 mm. Rolling plains have soils that are coarse-textured entic Haplustols with underlying calcareous hardpan and low organic matter as well as very sandy Entisols in dune areas. The natural vegetation is an open savanna grassland with mainly *Prosopis* tree species.
NE Brazil	Area is approximately 1,000,000 km², with tropical semiarid climate. Mean annual temperature is 25°C. Mean annual rainfall is 300–1,000 mm. Native vegetation is deciduous thorn forest. Soils are poorly developed and shallow (≈1 m) and highly variable spatially. Entisols and Alfisols are predominant on hilly land with only 5% floodplains.
Yucatán, Mexico	Area is approximately 40,000 km², with tropical semiarid climate. Mean temperature is 26°C. Precipitation gradient is from north to south, with 500–1,600 mm mean annual rainfall. The natural resource base is limited by geological and environmental conditions. Most of the peninsula is at or near sea level, with little elevation or relief, and 80% of the area is covered with flat layers of calcareous (limestone) materials of marine origin, with soil present in between this material. Soils are a highly variable mosaic of different types, especially of Rendolls, lithic Entisols, Alfisols, and Inceptisols, varying according to local conditions.

Key Findings

Shifts from native vegetation to pastures or agricultural fields led to severe decreases of soil fertility, carbon losses up to 60 percent, and significant erosion (Fraga and Salcedo 2004; Zach et al. 2006). Soil management alternatives to reduce soil erosion, such as vegetation strips (*Opuntia* sp.) planted on contour, were able to retain up to 30 tons of soil per hectare per year (Galindo et al. 2005).

Main Land Use Systems	Socioeconomic Characteristics
In the eastern, more humid region, use is mixed, with livestock and cash crops. In the western, drier region, most of the area is covered with natural savanna and is used for extensive cattle ranching. Deforestation is increasing, and cash crops are being introduced.	Area has 0.3 million inhabitants. Population density varies from 32 persons/km^2 in the capital to 0.1 persons/km^2 in the western ranching areas. Average farm size is 400 hectares in the east and over 1,000 hectares in the west. About 80% of land is privately owned; family labor is predominant. Annual per capita income depends on commodity prices but in small farms does not exceed US$ 10,000.
Use is for dryland subsistence agriculture (maize, beans, and cassava), livestock production based on native vegetation, and fuel wood harvest. Irrigation is practiced in less than 2% of the area because of lack of available water and/or suitable soil conditions.	Area has 25 million inhabitants. Population density varies from 10 to 30 persons/km^2. About 75% of farms have less than 10 hectares; 95% have less than 50 hectares. Practically all are privately owned; family labor is predominant. Annual per capita income is US$ 500.
Dominant land use systems in recent years have included shifting cultivation, sisal plantations, home gardens, cattle grazing systems, pig production systems, and secondary and primary forest. About 25% of Mexico's biosphere reserves are in the Yucatán Peninsula.	Yucatán State has 1.7 million inhabitants with a population density of 42 persons/km^2. About 51% of the population is active; 17% works in agriculture, 28% in industry, and 53% in services. Land tenure can be private or communal (ejido). Annual per capita income is US$ 1,200.

A positive relationship between soil clay plus silt content and soil organic matter proved to be an indicator of soil degradation and productive potential and has been used for land use planning (Quiroga et al. forthcoming). The increasing area of cash crop production in the humid areas of La Pampa has also caused higher stocking rates in the savanna ranching region, and overgrazing has caused shifts in plant species composition and loss of biodiversity (Frank et al. 2003; Noellemeyer et al. forthcoming).

The understanding of the relationship between crop demand and the quality of organic fertilizer available within the farms, nitrogen mineralization, and synchronization in northeast Brazil allowed us to suggest fertilization practices that increased crop productivity up to 300 percent (Menezes et al. 2002b; Silva and Menezes forthcoming). Management alternatives to improve soil fertility developed together with farmers included agroforestry, green and animal manures, and cover crops (Caamal et al. 2001; Castillo et al. 2003). These resulted in 130 percent increases in biomass productivity and increased soil organic matter and nutrient content by up to 40 percent (Menezes et al. 2002a; Tiessen et al. 2003; Perez and Menezes 2004; Perez et al. 2005).

In the Yucatán Peninsula, soil organic matter and nutrient levels were higher than in non-calcareous semiarid tropical soils elsewhere, and the rapid productivity decline in cropping fields was due to interactions between phosphorus and potassium supply and available water and also weed competition during the cropping cycle (Weisbach et al. 2002). We developed alternative weed management practices for the milpas (the traditional maize production system with intercrops of squashes, peppers, and others) in Yucatán, which helped to reduce herbicide use, increased maize productivity, and extended the cultivation period in the slash-and-burn cycle from two to four years. In addition to milpas, Mayan home gardens were an important component of the economy of rural families, providing fruit and vegetables, surpluses for the market, and sites for *in situ* conservation of agricultural diversity (Aké Gómez et al. 2002; Jiménez-Osornio et al. 2004; Querejeta et al. forthcoming).

Networking

A successful international network is built upon successful local and regional ones. This is particularly true when studying land use change and its environmental and socioeconomic impacts, because it requires an interdisciplinary and participatory approach in order to link researchers' and farmers' knowledge. This often requires institutional arrangements capable of dealing with issues of human welfare at scales from local to global levels.

Most research institutions in Latin America have relegated extension efforts largely to the level of secondary activities, although extension may well be an important link between universities and societies. Meanwhile, several nongovernmental organizations (NGOs) in developing countries have taken on extension roles and support the development of rural communities.

In northeast Brazil, a regional network was developed among two universities—the Universidade Federal de Pernambuco (UFPE) and the Universidade Federal da Paraíba (UFPB)—the NGO Assessoria e Serviços a Projetos em Agricultura Alternativa (ASPTA), and farmers' associations and local leaders. Several existing arrangements proved to be important for successful interaction between stakeholders. ASPTA had been working in the study region for six years prior to the beginning of our activities. For this and other reasons, ASPTA was already trusted by farmers. This provided the basis for the participatory identification and development of several research topics. ASPTA and farmers expected the research partners to find answers to management questions that the NGO technicians were not able to approach because it required scientific experiments, rather than demonstration plots. Before this alliance, research of UFPE and UFPB was mostly conducted at research stations with little connection to farmers and NGOs. The connection to ASPTA and farmers has redirected and refined some of the researchers' activities.

The establishment and maintenance of the interactions between these institutions in semiarid northeast Brazil was possible only due to long-term funding by several international agencies. The first contacts between UFPE scientists and ASPTA technicians were established in the mid 1990s during activities within a research project funded by the Kew Gardens, UK. Meanwhile research activities funded by the Canadian International Development Agency (CIDA) brought together scientists from the University of Saskatchewan (U of S), UFPE, and UFPB. Later Inter-American Institute for Global Change Research (IAI) funding allowed for a long-term intensive interaction between UFPE, UFPB, U of S, and ASPTA. This, in turn, provided the basis for additional Brazilian National Council of Technological and Scientific Development (CNPq) funding for research projects on family agriculture.

In La Pampa, central Argentina, the IAI project strengthened existing collaboration between the University of La Pampa (UNLP) and the Instituto Nacional de Tecnología Agropecuaria (INTA) on land management research. Again, long-term funding was the key to involving farmers, agricultural cooperatives, and NGOs such as the Asociación Argentina de Productores en Siembra Directa (AAPRESID) and the Asociación Argentina de Consorcios Regionales de Experimentación Agropecuaria (AACREA) in research activities. It was also crucial for integrating graduate and postgraduate students in the research. Most research was conducted on farmers' land, and research objectives were discussed and defined with farmers and their organizations. For example, farmers expressed interest in introducing cover crops, and we have since set up new experiments to determine the effects of fallow cover crops on crop yield, water availability, weed infestation, and carbon balance.

Our long-term tillage plots are installed on farms that belong to AAPRESID and AACREA and are widely used by these NGOs as demonstration plots during field days. This close collaboration over a relatively long period resulted in UNLP researchers' being regarded as the scientists who solve farmers' problems. Some of them are invited regularly to field days and informative talks, and so are many of the students who collaborated as part of their masters work. This interaction constantly generates new research demands by and interests from farmers, who offer their land for experimentation. In addition both NGOs contributed to research funding by providing inputs for on-farm experiments. The consolidation and acceptance by farmers and local scientists of the international research network work also had an impact on local government. The provincial government has asked UNLP scientists to study carbon sequestration and balance under native and cultivated forests. In addition the local government has asked for advice and consultations on the new provincial soil conservation law passed in 2004. The provincial secretary for the environment has made use of our project's publications to raise awareness about soil degradation in agricultural and natural ecosystems.

In Yucatán the network includes universities and NGOs from the region; the country; North, Central, and South America; the Caribbean; Europe; producer organizations; and the government. The group of the Manejo y Conservación de Recursos Naturales Tropicales (PROTROPICO) from the Universidad Autónoma de Yucatán (UADY) has played a role as facilitator for different actors and activities. A key point for PROTROPICO's success has been the long-term planning of multidisciplinary projects, integrating training and research and considering local problems. Strategic and applied research has been combined and sources of economic support have been available since 1992. PROTROPICO was already working in the region with support from the Ford, Rockefeller, and MacArthur Foundations when IAI involvement started. The support from these foundations has been maintained, and other national and international sources have contributed to PROTROPICO's projects. In addition we have helped NGOs and producers to obtain economic resources. This strategy has attracted students and researchers from different parts of Mexico and other countries, and PROTROPICO is now a reference center for the region.

Outreach

Little of the scientific information exchanged between scientists reaches the rural communities (Heong and Escalada 2005). However, our Collaborative Research Network (CRN) has developed the means for delivering scientific information to farmers and for supporting predictions at a regional scale.

In La Pampa our information is used for soil management recommendations as well as for land use policy decisions on a regional scale. Several bulletins of the Argentinean Agricultural Research Service (INTA) contain recommendations for soil management and conservation developed in the project (Quiroga et al. 2001; Vallejo et al. 2002; INTA 2004; INTA 2005). When the climate shifts toward drier conditions, marginal soils are susceptible to degradation because of water stress, lower yields, and excessive production pressure (Sierra et al. 2001). Regional diagnostics and strategic land use planning will help to reduce the impact of these changes. The tool for this strategic planning is a spatially explicit land use model that has been developed for agroecosystems in northeast Brazil and La Pampa and that takes into account different climate scenarios. Another important factor for land use decisions and policy is the increasing trend in La Pampa toward off-farm income and part-time farming, and the importance of non-rural capital funds that invest in cash crops. Therefore one of our activities has been to collaborate with economists, agronomists, and communications scientists who study the competitiveness of rural small and medium enterprises in the most important agrifood production and marketing chains.

The low-input land use systems of northeast Brazil have to rely on resource conservation and use efficiency in order to increase productivity and sustainability. Therefore we joined a multi-institutional participatory effort to develop management alternatives that include agroforestry systems and organic fertilization. For example, ASPTA had suggested the introduction of alley cropping systems with *Gliricidia sepium*, but the innovation did not bring the expected benefits and instead reduced crop yields. Our research identified strong competition for water between trees and crops. Based on this, we proposed the adoption of other agroforestry practices that avoid competitive interactions, such as live fences, planting of trees within areas unsuitable for agriculture, and the transfer of tree biomass to crop fields. We also tested native and introduced tree species that may be less competitive within simultaneous agroforestry systems.

While developing management practices for organic residues, our understanding of the relationship between the quality of organic fertilizer available within the farms, nitrogen mineralization, and synchronization to crop demand allowed us to suggest fertilization practices that led to significant increases of crop productivity. In addition we developed management techniques that help reduce erosion and runoff. Several farmers within the study region have implemented these practices on their land.

Research conducted in Yucatán has been evaluated by state and federal governments. The program developed by the Secretaria de Medio Ambiente Recursos Naturales y Pesca (SEMARNAT) for a permanent system of corn production was based on results of a network in which PROTROPICO was working. Two major projects that will have impact in the region are Ecological Planning of the State of Yucatan and the Forest Development Program of Yucatan State. These projects have been funded by the Mexican government; Fundacion Produce Yucatan, A.C.; and the Mesoamerican Corridor Project and are being coordinated by PROTROPICO. These plans will have a direct impact on land use of the region for the coming decades. In addition there is a network of home gardens in rural secondary schools that are used to reevaluate traditional knowledge and provide living laboratories and *in situ* conservation of agrobiodiversity. This is a joint project of PROTROPICO and the Ministry of Education that has received economic support from the Small Donation Programme of the UN Development Programme.

Capacity Building

Products of capacity building at the regional scale are about 90 students who were involved with our activities throughout the five-year period within the three study regions. Student interest was driven mostly by the diversity of a network involving research institutions, NGOs, and farmer organizations that permitted high-quality applied research in the fields of and in close contact with farmers. Some of the students came from farm families, taking advantage of the hands-on training in science. In northeast Brazil, from a total of 19 students, 6 were hired by regional NGOs, 6 were hired by research institutions (Embrapa, universities), and 5 went on to pursue graduate degrees. In La Pampa, 7 graduate students and more than 20 undergraduate students were supported by the project. Four of these students have been hired by INTA, 7 by AACREA and AAPRESID, and 2 by cash crop investment funds. In Yucatán 20 undergraduate and 25 graduate students were involved with our activities and now work in NGOs, other universities, the Yucatán government, and as consultants in the Mesoamerican Biological Corridor, while some have continued their training in other universities. Of the foreign graduate students, several work for universities in their countries or in research centers such as the Centro Agronómico Tropical de Investigacíon y Enseñanza (CATIE).

Challenges for the Future

Overall the project allowed for significant improvements in teaching, research, and extension activities in our institutions and throughout our study regions. Scientific training and involvement of students with NGOs and farmer associations has worked as a multiplying mechanism for high-quality and relevant work based on scientific data and local knowledge. In addition continued funding during the five-year CRN period has set the conditions for long-term interactions between stakeholders within diverse institutional arrangements, allowing for strong impact at the local scale (municipalities) and providing support for decision making at the regional scale.

One of the challenges for the future will be to bring continuing funding to maintain and expand research activities. Knowledge generated must also be used increasingly to influence broader regional policy formulation, in order to reach other areas and communities with characteristics similar to those within our study areas.

Literature Cited

Aké Gómez, A., M. Avila, and J. J. Jiménez-Osornio. 2002. Valor de los productos directos del agroecosistema solar: el caso de Hocabá, Yucatán, México. *Sociedades Rurales, Producción y Medio Ambiente* 3(1):7–18.

Caamal, A., J. J. Jiménez-Osornio, A. Torres-Barragan, and A. L. Anaya. 2001. The use of some allelopathic plants for weed control in agricultural management. *Journal of Agronomy* 93(1):27–36.

Castillo-Caamal, J. B., J. J. Jiménez-Osornio, A. López-Pérez, W. Aguilar Cordero, and A. M. Castillo-Caamal. 2003. Feeding mucuna beans to small ruminants of Mayan farmers in the Yucatán península, México. *Tropical and Subtropical Agroecosystems* 1(2–3):113–7.

Fraga, V. S., and I. H. Salcedo. 2004. Declines of organic nutrient pools in tropical semi-arid soils under subsistence farming. *Soil Science Society of America Journal* 68:215–24.

Frank, F., E. Frank, and E. Noellemeyer. 2003. *Soil texture and carbon relationships under different managements in the caldenal area of La Pampa, Argentina.* Ribeirão Preto, São Paulo, XXIX Congresso Brasileiro de Ciência do Solo. Viçosa: Sociedade Brasileira de Ciência de Solo.

Galindo, I. C. L., E. V. S. B. Sampaio, and R. S. C. Menezes. 2005. Uso da palma na conservação dos solos. In Menezes, R. S. C., D. A. Simões, and E. V. S. B. Sampaio, eds. *A Palma no Nordeste do Brasil: conhecimento atual e novas perspectivas de uso.* Recife: Editora Universitária Universidade Federal de Pernambuco (UFPE). (pp. 163–176)

Heong, K. L., and M. M. Escalada. 2005. Scaling up scientific information to rural communities. *Journal of Science Communication* 4(3):1–3.

INTA EEA Anguil. 2004. Producción y calidad de verdeos de invierno. *Boletín de divulgación técnica* No. 80. Anguil, La Pampa: Argentinean Agricultural Research Service (INTA), Estación Experimental Agropecuaria (EEA).

INTA EEA Anguil. 2005. Aspectos de manejo de los suelos en sistemas mixtos de las regiones semiárida y subhúmeda pampeana. *Boletín de divulgación técnica* No. 87. Anguil, La Pampa: Argentinean Agricultural Research Service (INTA) Estación Experimental Agropecuaria (EEA).

Jiménez-Osornio, J. J., M. R. Ruenes-Morales, and A. Aké Gómez. 2004. Mayan home gardens: Sites for *in situ* conservation of agricultural diversity. In Jarvis, D. I., R. Sevilla-Panizo, J. L. Chavez-Servia, and T. Hodkin, eds. *Seed Systems and Crop Genetic Diversity On-Farm. Proceedings of a Workshop.* Rome: IPGRIP. (pp. 9–15)

Menezes, R. S. C., E. V. S. B. Sampaio, L. M. Silveira, H. Tiessen, and I. H. Salcedo. 2002a. Produção de batatinha com incorporação de esterco e/ou crotalária no Agreste paraibano. In Silveira, L., P. Petersen, and E. Sabourin, eds. *Agricultura familiar e agroecologia no semi-árido: avanços a partir do agreste da Paraíba.* Rio de Janeiro: Assessoria e Serviços a Projetos em Agricultura Alternativa (ASPTA). (pp. 261–70)

Menezes, R.S.C., I. H. Salcedo, and E. T. Elliott. 2002b. Microclimate and nutrient dynamics in a silvopastoral system of semiarid northeastern Brazil. *Agroforestry Systems* 56(1):27–38.

Noellemeyer, E., D. Estelrich, and A. Quiroga. Forthcoming. Soil quality in three range soils of the semiarid Pampa of Argentina. *Journal of Arid Environments.*

Perez, A. M., and R. S. C. Menezes. 2004. Umidade ao longo do perfil do solo em um sistema agroflorestal com gliricídia e milho. In *XV Reunião Brasileira de Manejo e Conservação do Solo e da Água,* Santa Maria, RS. Viçosa: Sociedade Brasileira de Ciência de Solo.

Perez, A. M. M., R. S. C. Menezes, and E. D. Silva. 2005. Nutrientes do solo, microclima e produtividade do milho em sistema agroflorestal com *Gliricidia sepium* no Agreste paraibano. *Revista Brasileira de Ciência do Solo* 30(3):555–64.

Querejeta, J. I., H. Estrada, M. F. Allen, J. J. Jiménez-Osornio, and R. Ruenes. (Forthcoming)a. Water source partitioning among Mexican native trees growing on shallow soils over karst in a seasonally dry tropical climate. *Oecologia.*

Quiroga, A., D. Funaro, E. Noellemeyer, D. Peinemann. (Forthcoming)b. Soil quality indicators and barley fertilization response. *Soil & Tillage Research.*

Quiroga, A., O. Ormeño, and N. Peinemann. 2001. Materia orgánica. Un indicador de la calidad de suelos relacionado con la productividad de los cultivos. *Boletín de divulgación técnica* No. 70. Anguil, La Pampa: Argentinean Agricultural Research Service (INTA), Estación Experimental Agropecuaria (EEA).

Sierra, E., S. Perez, G. Casagrande, G. Vergara. 2001. Efectos del ENSO sobre las precipitaciones del trimestre noviembre-enero (1921/1998) en el centro-este de la provincia de La Pampa (Argentina). *Revista Argentina de Agrometeorología* 1:83–7.

Silva, T. O., and R. S. C. Menezes. Forthcoming. Adubação orgânica com esterco e/ou

Crotalaria juncea: disponibilidade de N, P e K ao longo do ciclo de cultivo. *Revista Brasileira de Ciência do Solo.*

Tiessen, H., R. S. C. Menezes, I. H. Salcedo, and B. Wick. 2003. Organic matter transformations and soil fertility in a treed pasture in semiarid northeast Brazil. *Plant and Soil* 252(2):195–205.

Vallejo, A., R. Souto, and A. Quiroga. 2002. Siembra directa y ferilización en sistemas ganaderos de la región semiárida. *Boletín de divulgación técnica* No. 74. Anguil, La Pampa: Argentinean Agricultural Research Service (INTA), Estación Experimental Agropecuaria (EEA).

Weisbach, C., H. Tiessen, and J. J. Jiménez-Osornio. 2002. Soil fertility during shifting cultivation in the tropical karst soils of Yucatan. *Agronomie* 22:253–63.

Zach, A., H. Tiessen, and E. Noellemeyer. 2006. Carbon turnover and ^{13}C natural abundance under landuse change in the semiarid La Pampa, Argentina. *Soil Science Society of America Journal* 70:1541–6.

17

Cattle Ranching, Land Use, and Deforestation in Brazil, Ecuador, and Peru

Charles H. Wood and Jean-François Tourrand

A key dimension of global environmental change is the high rate of deforestation taking place in the lowland tropics of Central and South America, where landholders clear large tracts of primary and secondary forests to establish pastures for raising cattle (Durning and Brough 1991). Deforestation is further propelled by the unsustainable pasture management technologies landholders often use. The result is pasture degradation that compels ranchers to clear additional forest to maintain their herds.

So rapid is the expansion of cattle ranching in the Brazilian Amazon that concerned analysts coined the term *pecuarização* (cattlelization) to depict the extraordinary increase in cattle herds among farmers and ranchers, both large and small. The expansion of cattle ranching is also underway in Peru and Ecuador, although more slowly and in response to different social, economic, and political incentives. Given the close relationship between cattle ranching and deforestation, a study of the factors that motivate farmers to become ranchers and an analysis of the choices landholders make with respect to pasture management are effective ways to address one of the paramount environmental issues of our times.

Comparative Design

The differences between Brazil, Peru, and Ecuador called for a comparative design capable of generating insights and conclusions not captured by the more common focus on a single country or a particular region within a country. In Peru the focus was on the Alto Huallaga (*selva alta*) and Aguaytia–San Alejandro (*selva baja*) regions. In Ecuador the focus was on the Valle de Quijos (*selva alta*), the Puyo–Santa Clara–Palora (*piedemonte*) area, and the Coca–Francisco de Orelanna (*selva baja)* region. In

Brazil the analyses targeted the eastern, central, and western regions, each of which displays more or less distinct characteristics. Underlying this approach was the assumption that a comparison of results from different countries and from varied regions within Brazil could offer new insight into the causes of the expansion of the cattle population in the Amazon.

Land Use Decisions

Varied perspectives concerning the causes underlying the growth in cattle ranching vie for prominence in the environmental literature. A once-influential school of thought maintained that raising cattle is not economically profitable, per se, and that its growth in the Brazilian Amazon was mainly due to factors such as government subsidies and land price speculation (Hecht 1992). A more recent perspective contends that the increase in cattle ranching is economically profitable in Brazil due to the development of regional economies and increases in urban demand (Faminow 1998). A third point of view emphasizes the intrinsic advantages of cattle ranching, especially to small farmers, compared with other agricultural activities (Smith et al. 1995; Durning and Brough 1991).

Because the policy implications that arise from the opposing conclusions are different, the debate regarding the causes of increased cattle ranching has assumed paramount importance (Andersen et al. 2002). In light of the significance of these issues, it was necessary to go beyond general perspectives to achieve a more complete understanding of the social and ecological factors that account for the pace and pattern of land use change.

To address the complexity of the causal relationships at hand, we chose an interdisciplinary research strategy that targeted the decision-making process by which firms and households allocate their resources, including land, labor, and capital. Figure 17.1 highlights the three "critical decisions" that served to orient the process of data collection and analysis: 1) the decision to invest in cattle, 2) the choice of pasture management strategies, and 3) decisions regarding the clearing of forested areas.

Having placed key land use decisions at the center of the analytical focus, we then constructed a research design that investigated how the three critical decisions observed at the local level were, in turn, influenced by institutions, events, and decisions at regional, national, and international levels. The research objective was thus to comprehend not only the micro decision processes made by farmer/ranchers but also the decision processes made by stakeholders at successively more distant levels of social organization, at the meso and macro scales. By specifying the constellation of forces and processes that operated at different analytical scales, we sought to identify critical points in the causal chain of relationships where policy recommendations could reduce the environmental damages commonly associated with cattle ranching.

Further analysis focused on the stages that link the birth of a calf to the final con-

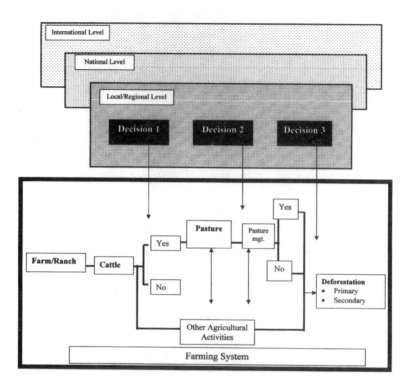

Figure 17.1. Decision process for land use choices

sumption of beef. Links in the "marketing chain" are defined by specific activities, such as the purchase, transport, fattening, and slaughter of animals. Each activity along the way involves a set of more or less distinct social actors whose decisions, like those made by ranchers, are conditioned by local/regional, national, and global level considerations. Attention to marketing chains proved important because marketing chains determine the prices that ranchers confront as well as the quality standards ranchers are required to meet.

Data Collection: Stakeholders as Participants

In-depth interviews with land managers were carried out by interdisciplinary teams of researchers composed of experienced field-workers from the United States, Brazil, Peru, Ecuador, and Canada. Rather than apply a formal questionnaire to hundreds of randomly selected respondents, as in survey research, the research teams carried out "interactive interviews" with a small number of systematically chosen key informants. Key informants were individuals who occupied a central position in the system and therefore possessed relevant knowledge and information. The interactive interview is like a "directed conversation" that allows the informant the freedom to respond in different ways. Questions, prompts, comments, and clarifications may direct informants to particular topics and invite them to share the knowledge derived from their own lived experience. Respondents are allowed to determine the relevance of different issues via a strategy that is seemingly improvisational yet is focused on project goals. When the participatory interview is properly conducted, the informant becomes actively involved in the production of knowledge (Gubrium and Holstein 1997). The conversation allows stakeholders to introduce topics that were often not anticipated by the interviewer (or existing theory), thereby allowing for the possibility of surprises and novel insights.

Interdisciplinarity and the Capacity to Hear Stakeholders' Views

The interviews with key informants and stakeholders were carried out by teams of researchers who represented a wide range of disciplinary specializations, including animal science, veterinary medicine, farming systems, economics, sociology, anthropology, geography, and political science.

A useful innovation was to have Brazilian researchers participate in the interviews carried out in Peru and Ecuador, and vice versa, as a means to offset the common tendency for researchers to assume that what they know about their own country equally applies to others. The mixing of interviewers from different nationalities also contributed to the comparative analysis by sensitizing researchers to similarities and differences that would otherwise have been overlooked.

Linking Science and Policy

The causes of land use change in the Amazon cannot be reduced to a single variable or to an isolated process. History, politics, and demography are implicated, as are exchange rates, monetary inflation, legal institutions, road construction, colonization schemes, tax laws, financial markets, commodity prices, and tenure security, to name the more salient variables noted in the literature. It is further recognized that the biophysical context—defined by such variables as soil quality, water availability, altitude and temperature range, and the presence of pests and pathogens—mediates the way that socioeconomic factors play themselves out in a particular location. The image that emerges from these considerations is that of a complex web of relationships that operate within and across different levels of social and ecological organization.

The multileveled approach depicted in Figure 17.1 was one way to deal with these complexities. Rather than attempting to hold all else constant and to isolate the individual effects of particular variables, as in a statistical regression model, this approach allowed for the multiple variables and processes that result in expansion of cattle ranching. The project thus attempted to understand the decision-making process that took place within rural households and agricultural firms but also treated those decisions as processes embedded in larger systems. The approach was therefore both reductionist and holistic, more or less in the manner endorsed by hierarchy theory in landscape ecology (Ahl and Allen 1996; Allen and Starr 1982).

The methods of data collection, as well as the larger design of the project, were conducive to narrowing the gap that often separates scientific analysis and policy formulation. By including stakeholders as participants in the interactive interviews of key informants, the findings necessarily incorporated the interests, aims, and attitudes of important players. Moreover, the multileveled design positioned the various stakeholders' interests within the larger constellation of processes that operated at various scales. As a result, we were sometimes in a position to locate those critical points in the causal chain of relationships that were responsive to policy change.

Capacity Building

The project contributed to capacity building by including young scholars in fieldwork and by providing modest research awards for carrying out independent analyses of relevant topics. Direct or in-kind support played a part in 20 academic degrees that benefited graduate and undergraduate students in a variety of disciplines, including anthropology, ecology, economics, sociology, geography, agrarian science, agroforestry, political science, and agronomy.

Conclusions

- In recent years, the size of the cattle population in the Brazilian Amazon has increased by around 5 percent a year, an estimate that is substantially higher than the comparable rate of growth in Peru and Ecuador.
- Ranching was once considered uneconomical in the absence of fiscal incentives that were captured mainly by large landholders. Today ranching can be profitable for both small and large landholders for many reasons, including the low cost of production and the ability to supply markets outside the region.
- Beef produced in the Amazon is increasingly sold in the Brazilian northeast and in the south, where producers divert higher-quality beef to international markets.
- Cattle ranching is an attractive option for small landholders for many reasons: Cattle are a highly liquid asset; beef prices have been stable compared with other agricultural commodities; labor needs are relatively low and can be met by members of the household; beef and milk production complement each other and can be integrated into the other agricultural activities; pastures can be established at low cost in areas initially cleared for the cultivation of food crops; fence posts and related mate-

rials are often provided by sawmills in exchange for timber; and annually setting fire to the pasture is a low-cost, low-skill pasture management technology.

- Compared with other places in Brazil, such as the northeast, the biophysical properties of the Amazon are favorable for cattle ranching. Pasture grasses such as *Brachiaria brizantha* and selected breeds of cattle such as Zebu and mixed Zebu/European breeds are generally well adapted to the region.
- Cultural considerations enter into the decision to invest in cattle to the extent that ranching activities validate desired social identities.
- The choice of pasture management strategies depends on the internal characteristics of rural households, on price and market structures, and on the historical and cultural characteristics of the population.
- Household consumption of milk is giving way to low-cost production of dairy products sold in urban areas in the region and, increasingly, in other parts of the country.
- The growth, profitability, and quality standards of beef and milk production are strongly influenced by the presence and character of marketing chains.
- Federal credit policies intended to promote agriculture in Brazil provided an incentive for small landholders to invest in cattle. Credit policies have had less influence in Peru and Ecuador.
- In Peru the expansion of cattle ranching may have reduced deforestation in areas once devoted to coca production. Rather than clear additional forest, landholders establish pastures in former coca fields, where the depleted soils are unsuitable for the cultivation of perennial crops.

Deforestation was associated with cattle ranching in all of the research sites, although the magnitude of deforestation and the specific combination of factors that led to land cover change varied from one context to another and varied by type of producer. Whatever the circumstances, investing in cattle remains an attractive option for small and large landholders, who have demonstrated a remarkable capacity to adapt to new environments and changing socioeconomic conditions. The evidence suggests that cattle ranching will continue to expand in years to come. The interdisciplinary and multileveled approach adopted in this project is one way to identify and target those points in the "architecture" of shaping influences that are amenable to policy initiatives designed to slow deforestation and promote sustainable pasture management in the Andean and Brazilian Amazon.

Literature Cited

Ahl, V., and T. F. H. Allen. 1996. *Hierarchy theory: A vision, vocabulary and epistemology.* New York: Columbia University Press.

Allen, T. F. H., and T. B. Starr. 1982. *Hierarchy: Perspectives for ecological complexity.* Chicago: University of Chicago Press.

Andersen, L. E., C. W. J. Granger, E. J. Reis, D. Weinhold, and S. Wunder. 2002. *The dynamics of deforestation and economic growth in the Brazilian Amazon.* New York: Cambridge University Press.

Durning, A., and H. Brough. 1991. *Taking stock: Animal farming and the environment.* Worldwatch Paper 103. Washington, D.C.: Worldwatch Institute.

Faminow, M. D. 1998. *Cattle, deforestation and development in the Amazon: An economic, agronomic and environmental perspective.* CAB International.

Gubrium, J. F., and J. A. Holstein. 1997. *The new language of the qualitative method.* New York: Oxford University Press.

Hecht, S. B. 1992. Logics of livestock and deforestation: The case of Amazonia. In Downing, T. E., S. B. Hecht, H. A. Pearson, and C. Garcia-Downing, eds. *Development or destruction: The conversion of tropical forests to pasture in Latin America.* Boulder, Colo.: Westview Press. (pp. 7–25)

Smith, N., E. A. Serrão, P. T. Alvim, and I. Falesi. 1995. *Amazonia: Resiliency and dynamism of the land and its people.* New York: United Nations Press.

18

Global Change Effects on the Vegetation of Tropical High Mountains and Savannas

Carlos A. Klink, Juan F. Silva, Aura Azócar, Juan González, and Ricardo Herrera-Peraza

How do savannas and tropical mountain ecosystems respond to global change? What are the main changes in structure and functioning of these ecosystems when subjected to human modification and use? What are the ecological feedbacks? To answer these questions, we used two lines of research: analyses of changes taking place along natural and "human disturbance" gradients and analyses of ecological structure and dynamics at scales from landscape to organisms. This approach allowed scientists to shed light on how ecosystems or their components have responded to a range of environmental changes.

Some of this scientific information can be readily used in decision-making processes. For instance, our results were influential in showing the effects of large-scale transformation of natural landscapes into pastureland, from both Andean cloud forests and native savannas. Andean natural vegetation exerts a strong moderating effect on surface runoff, particularly during major rainfall events. This feature is lost when the natural vegetation is replaced by pastures. Floristic composition affects the magnitude of surface drainage, which in Colombian pasturelands reaches very high values (Ataroff and Silva 2005). The regulating effects of forests determine the magnitude of downstream flows and reduce the occurrence of flooding.

The scientists in our project translated scientific information for different audiences in order to influence decision-making processes. We were able to apply scientific information to regional land use decisions based on the balance between human needs and ecosystem function (Klink et al. 2005). The ongoing, rapid transitions to pasture and cropland taking place in the Cerrado savannas of Brazil have multiple consequences for biodiversity, watershed protection, carbon sequestration, and other ecosystem functions.

Box 18.1. Key findings

Cloud forests of the Andes capture atmospheric water and regulate flux and drainage from 3,000 m to the piedmont. Fragmentation of these forests, or their transformation into planted grasslands, may affect water flow, flooding, and landslides.

There is evidence of climatic change in the lowland of the Province of Tucumán, Argentina, with increased average annual rainfall, more intense rainfall events, and increased cloudiness, which causes lower day temperatures and higher night temperatures. This combination of changes has favored the transformation of natural ecosystems into agricultural systems.

Many of the dominant plant species of tropical high mountains are specialists and respond similarly to thermal and water stresses. Under climate change, they are likely to be replaced by fast-growing species with broad ecological niches. The most-affected species would be caulescent rosettes and shrubs. Avoidance is the main resistance mechanism to low temperature of the dominant rosette species. Low-growing plant species (cushion plants, grasses, herbs, and acaulescent rosettes) rely on freezing tolerance. Regarding water stress, species lie along an avoidance-tolerance gradient: Most shrubs and caulescent rosettes are avoidance types, while herbs and grasses are tolerant.

The genus *Polylepis* is a unique ecological model for the study of plant distribution along altitudinal gradients because it reaches much higher elevations than any other angiosperm tree. *P. sericea* has a wide altitudinal/latitudinal distribution range; therefore populations from low altitudes could replace those from high altitudes. *P. australis* is restricted to extreme southern latitudes; therefore populations would not be replaced. Under climate change, the genus would persist, while species would be affected differently.

Drainage, topography, and dominant vegetation are the main factors to discriminate landscape and ecological units of the Cerrado, Brazil. Some edaphic features differ at the landscape scale but are less important in differentiating ecological units.

Landforms and soil types affect woody vegetation of the Llanos, Venezuela. Tree

(continued on next page)

Therefore, the trade-offs between economically valuable production of beef, grain, and other agricultural products could be compared with multiple ecosystem functions over various temporal and spatial scales (Klink et al. 2005).

Communicating these trade-offs influenced how decision makers used scientific knowledge, for instance, in a white paper commissioned by the World Bank in support of the Sustainable Cerrado Initiative to be financed by the Global Environment Facility (Klink 2004) . In July 2005, this document became the basis for developing the Sus-

density is higher on sandy and more-stable surfaces despite lower water retention capacity in the topsoil. The increase of forest patches in the Orinoco Llanos in recent decades is not only the outcome of fire and grazing exclusion. Ecological processes such as establishment of new woody patches in the grass matrix, coalescence of neighboring patches, and persistence of established patches may be related to changes in rainfall patterns: increased monthly averages in June and July (early wet season) and decreased monthly averages from October through December (transition to early dry season). This trend may have caused wetter growth seasons and drier transitions from wet to dry seasons. Seedlings can produce enough biomass during their first season to allow them to resprout and survive after a dry season fire in the Llanos. This tolerance to fire indicates that water deficit (in space and time) is more relevant to the regeneration of woody species.

Deep roots affect the hydrology of the Cerrado. During the dry season, deep soil compartments can contribute as much as 80 percent of the total water used by tree-dominated communities. When there is below-average rainfall during the growing season, deep roots seem to provide water for the vegetation. Replacing natural ecosystems with planted pastures may change the water balance of the Cerrado. There is evidence of biological activity and availability of phosphorus in deep Cerrado soil. It seems that phosphorus supply is dependent on the biological cycling.

The range of natural abundance of nitrogen-15 in Cerrado plant leaves is similar to that in other nitrogen-limited ecosystems. The variability in the abundance of nitrogen-15 found is due to the presence of nitrogen-fixing legumes, associations with mycorrhiza, depth-related variability of soil organic matter, the seasonality of nitrogen immobilization and mineralization in soils, and fire.

Carbon stocks in the natural Cerrado are larger than in planted pastures (which cover 25 percent of the Cerrado). In both ecosystems, most carbon stocks from biomass are underground, where fluxes are more intense. Carbon recycling is faster in pastures. Management practices can change the direction of transfer (from land to atmosphere or from atmosphere to land).

tainable Cerrado program by the Brazilian Ministry of the Environment. The results of other initiatives that have implications for policy, training, and outreach related to global change are in the form of synthesis of results, such as the following: *Cambios globales, biodiversidad y el funcionamiento de ecosistemas montanos y sabanas en Sudamerica* (Silva et al. forthcoming); *Ecofisiología de plantas de páramo* (Azócar and Rada 2006); *Dinámica hídrica en sistemas neotropicales* (Ataroff and Silva 2005); and *Symposium: Cerrado Land-Use and Conservation: Assessing Trade-Offs Between Human and Ecological*

Needs (Klink et al. 2005). Still others are in the form of databases: "Plantas vasculares de los paramos de Venezuela: botánica y ecología" and "Plantas vasculares de las sabanas de Venezuela: botánica y ecología" (Silva and Ramirez forthcoming).

The proximity of scientists to practitioners and policy makers in environmental change issues can be positive for two reasons: Policy makers can incorporate scientific knowledge into their decision-making process, and scientists can guide their research toward problem solving. The key factor for a positive interaction is the willingness of both partners to participate in a dialogue. However, this is not an easy dialogue, because of the difficulties in understanding jargon and approaches, differences in time perspectives (particularly when practitioners ask scientists to put knowledge into use right away and scientists usually are concerned about whether or not results are "ready" for use), and the occasional perception that scientists are researching themes or subjects that are not relevant to what policy makers need.

One area in which the dialogue can be very fruitful is capacity building, because both arenas recognize and prize this endeavor. Our network focused on training young ecologists in the region pursuing South-South collaboration (Silva 2005). The network provided methodological and theoretical training for over 45 students and professors through exchange programs, courses, and workshops. Two professionals who received their doctorates in the network now work in the Brazilian Ministry of the Environment (on international programs in conservation) and the Ministry of Science (on climate change policies). Most exchange and training programs were associated with graduate programs within South America. This shows the maturity level of some academic centers in the region. Working in a Collaborative Research Network also helped to compensate for any weaknesses of the graduate programs involved. Students developed advanced research that was well integrated into the scientific program at relatively low cost. The effectiveness of the network is demonstrated by the number of scientific contributions made, most of them authored or coauthored by the students: 44 publications, 18 papers submitted, 13 articles in preparation, and 113 presentations at scientific meetings (Silva 2005).

The network also participated in outreach activities to link its scientific findings to the policy arena and society at large. Although we did not make systematic attempts to translate all of our results for a policy audience, some of our publications were translated into Portuguese and Spanish with that aim (Klink 2001; Azócar and Fariñas 2003; Silva 2003; Silva et al. 2003; Ataroff and Silva 2005; Minetti and González 2005).

What stands in the way of a better dialogue between scientists and policy makers? There are probably many responses to this question. The answers revolve around how to accommodate the dialogue into the strategic framework, how to compete for the best personnel for the job, and how to fund this as a long-lasting activity. To promote the dialogue between scientific disciplines and nonacademic arenas on environmental problems, it is crucial to identify instances of "ecological viability" that both transform our practice toward environmental change and help to find solutions to complex problems collectively (Agrawal 2005; Norgaard and Baer 2005). Therefore we suggest a more proactive role for the Inter-American Institute for Global Change Research and other funding agencies in this process, with scientists fulfilling the role of networkers with a full-time dedication and a well-developed agenda.

Acknowledgments

We would like to thank Charles Wood, Jerry Melillo, and Sônia Gianesella for comments and corrections in the manuscript.

Literature Cited

Agrawal, A. 2005. *Environmentality: Technologies of government and the making of subjects.* New Ecologies for the 21st Century Series. Durham, N.C.: Duke University Press.

Ataroff, M., and J. F. Silva, eds. 2005. *Dinámica hídrica en sistemas neotropicales. Investigaciones en Dinámica Hídrica de la red RICAS.* Mérida, Venezuela: Instituto de Ciencias Ambientales y Ecológicas (ICAE). (48 pp.)

Azócar, A., and M. Fariñas. 2003. Páramos. In Aguilera, M., A. Azócar, A. E. González-Jimenez, eds. *Biodiversidad en Venezuela, Tomo II.* Caracas, Venezuela: Ediciones Fundación Polar y Fonacit. (pp . 716–33)

Azócar, A., and F. Rada. 2006. *Ecofisiología de plantas de páramo.* Mérida, Venezuela: Instituto de Ciencias Ambientales y Ecológicas (ICAE). (178 pp.)

Klink, C. A. 2001. O papel da pesquisa ecológica na gestão ambiental e manejo de ecossistemas. In Burstyn, M., ed. *A Difícil Sustentabilidade.* Rio de Janeiro: Garamond Universitária. (pp. 77–84)

Klink, C. A. 2004. *Synthesis report for the Brazil Sustainable Cerrado Initiative.* Washington, D.C.: World Bank.

Klink, C. A., R. DeFries, and R. Cavalcanti, eds. 2005. *Symposium: Cerrado Land-Use and Conservation: Assessing Trade-Offs Between Human and Ecological Needs.* 19th Annual Meeting of the Society for Conservation Biology, Brasilia, Brazil, July 15–19, 2005. Discussion drafts of papers published in CD-ROM. Washington, D.C.: Universidade de Brasilia and Conservation International.

Minetti, J. L, and J. A. González. 2005 (submitted). El cambio climático en la provincia de Tucumán. Sus impactos. *Serie Conservación de la Naturaleza (Tucumán, Argentina).*

Norgaard, R. B., and P. Baer. 2005. Collectively seeing complex systems: The nature of the problem. *BioScience* 55(11):953–60.

Silva, J. F., A. Azócar, C. A. Klink, J. González, and R. Herrera-Peraza. 2003. Ecologia Tropical Para El Cambio Global. In *Inter American Institute for Global Change Research Annual Report 2002–2003.* Sao José dos Campos, Brazil: Inter-American Institute for Global Change Research (IAI). (pp. 20–33)

Silva, J. F. 2003. Sabanas. In Aguilera, M., A. Azócar, A. E. González-Jimenez, eds. *Biodiversidad en Venezuela, Tomo II.* Caracas, Venezuela: Ediciones Fundación Polar y Fonacit. (pp. 678–95)

Silva, J. F. 2005. Comparative study of the effects of global change on the vegetation of two ecosystems: High tropical mountain and tropical savanna. In *CRN-040 Final*

Report to IAI. Mérida, Venezuela: Instituto de Ciencias Ambientales y Ecológicas (ICAE).

Silva, J. F., and M. E. Ramirez. Forthcoming. *Plantas vasculares de las sabanas de Venezuela: botánica y ecología.* Mérida, Venezuela: Instituto de Ciencias Ambientales y Ecológicas (ICAE).

Silva, J. F., M. Ataroff, A. Azócar, M. Fariñas, C. Garcia-Nuñez, J. González, C. A. Klink, and F. Rada. Forthcoming. *Cambios globales, biodiversidad y el funcionamiento de ecosistemas montanos y sabanas en Sudamerica.* Mérida, Venezuela: Instituto de Ciencias Ambientales y Ecológicas (ICAE).

19

Linking Global Change Research to Improved Policies and Management for Amazonian Rivers

Michael E. McClain, Remigio Galárraga-Sánchez, Carlos A. Llerena, and José Efrain Ruiz

Science Framework

Forces of global change on the eastern slope of the Andes pose formidable challenges to sustainable development. Poverty is widespread in the Amazon portions of Colombia, Ecuador, Peru, and Bolivia, and growing population pressures are driving a process of rapid land use change that is largely uncoordinated and uncontrolled (Mena et al. 2006). Moreover this conversion of forests to mixed agriculture and pasture is expanding across one of the most biodiverse regions of the planet, and there are well-justified fears of significant biodiversity losses (Mittermeier et al. 1997). Predictions of potential climate change are not reliable in this region, but based on experience with intense climate variability linked to El Niño, Andean governments worry that climate changes will only exacerbate their current development problems (CONAM 2002). Rivers are fundamental in shaping and maintaining Andean Amazon ecosystems, and they are vital to the region's hopes for sustainable development and improved livelihoods of its inhabitants.

Rivers are integral to the health, welfare, and prosperity of human inhabitants and natural ecosystems of the Andean Amazon region, and impacts brought by environmental change threaten the services these rivers provide. Rivers are sources of the fish and semiaquatic animals in the diets of Amazon indigenous people and colonists, accounting for more than 50 percent of animal protein consumed in many areas (Galárraga et al. 2001; McClain et al. 2001; Ayllon 2002). River valley bottoms offer the only high-quality arable lands in the highlands, and at lower elevations, river banks offer the only high-fertility soil, supporting protein-rich crops that cannot be grown on upland soils (Cossio 2002; McClain and Cossio 2003). Rivers and streams are the main sources of drinking water, but they are also common receptacles for wastes (Galárraga et al. 2001; Puentes 2004; Blanco 2005). Consequently inhabitants of the region suffer from a wide array of waterborne illnesses, and diarrhea is a common ailment among children (McClain et al. 2001). In addition to the impacts of towns and cities, other threats linked to agriculture, cattle raising, forestry, and petroleum extraction also have measurable impacts on the condition and quality of rivers in the region (McClain 2002; Celi 2005). Rivers are vitally important to commerce; they serve as the primary avenue for transporting products to and from many portions of the region. Finally, rivers are an important part of people's daily recreation, as many families, and especially children, seek relief from the heat in the river's cool waters. For indigenous people, rivers also hold important spiritual significance.

The Andean Amazon Rivers Analysis and Monitoring (AARAM) project was financed by the Collaborative Research Network (CRN) program of the Inter-American Institute for Global Change Research (IAI) to examine the relationships between climate, terrain features, land use, human demographics, and river systems in the Andean Amazon region. The primary goal of AARAM was to provide scientific guidance for policy and management actions involving rivers and water resources, and the key science findings are summarized below. Work focused on characterizing the current spatial and temporal patterns in water quantity and quality and investigating functional relationships between quantity, quality, natural features, and human actions. The principal investigators of AARAM were all limnologists and hydrologists, and expertise in the human dimensions of resource management was sought through the involvement of local collaborators.

Key Findings

- Land use conversion is progressing rapidly across the Andean Amazon region in proportion to growing population densities and expansion of roads.
- Land cover configuration in areas controlled by colonists and indigenous groups is more correlated to biophysical variables limiting land uses than to cultural variables.
- Concentrations and fluxes of sediments and solutes from Andean catchments are correlated to both discharge and land use. Concentrations of sediments and nutrients increase with increasing discharge over daily time scales, while longer-term average concentrations increase with the proportion of agriculture and pasture upstream.
- While sediment concentrations become very high during large runoff events, solute nutrient and organic matter concentrations remain low as compared with more intensively used landscapes.
- The responsiveness of sediment and solute concentrations to daily runoff suggest, however, that Andean Amazon river systems are especially vulnerable to biogeo-

chemical changes in response to land use change, and more intensive future land use may produce strong changes in riverine biogeochemistry.

- People of the region rely on rivers for many services. The rivers are the primary source of water for all human uses and also a common receptacle for wastes. Fish are an important component of the local diet, and riverside soils produce the most nutritious crops.

- Andean rivers appear to be able to assimilate domestic wastes from riverside towns quite effectively, but the cumulative effects of these discharges significantly disturb riverbed habitats, even when contamination is difficult to detect in the water column.

- Management of water and riparian ecosystems is inconsistent across the region, as laws are loosely adhered to and enforcement is ineffectual. Household level compliance with management prescriptions was highest among indigenous people because of stronger community structure and respect for community authorities.

Relevance: The Political and Legal Situation

Despite the obvious value of rivers to the welfare and development potential of communities in the Andean Amazon, legal protection and careful management of rivers are not high priorities for national and local authorities. Rivers may, in fact, be the least protected and most vulnerable resources in the Andean Amazon. The explanation for this situation is multifaceted and differs somewhat from one part of the region to another, but certain common factors stand out.

From a national perspective, low population densities, relative poverty, and remoteness result in the Amazon's being commonly left out of the national dialogue on issues of public policy, legal reform, and community level investments for development. Another key consideration at the national level is the distribution and commercial value of resources in these countries. For example, petroleum and timber are regarded as resources of national importance in the Amazon. Thus the Amazon is the focus of considerable national attention from the ministries and agencies managing these resources, and national policies and regulatory agencies are in place—although regulatory agencies are not always effective at enforcement. Fisheries are regarded primarily as a resource of the coasts, and virtually all policies and investments are focused on the coast. Inland fisheries receive much less attention, and the little attention they get is focused on the big-river fisheries in the vicinity of major Amazon cities (Leticia, Iquitos, Pucallpa). Smaller artisan fisheries are generally unregulated. The legal and regulatory systems for water resource management are biased toward the semiarid and arid portions of Andean countries. Water management is thus viewed almost exclusively as water supply for irrigation, and laws and policies strongly reflect this view. We believe this to be a fundamental obstacle to effective river management in the Andean Amazon (McClain and Llerena 1998). As currently practiced, water management is primarily for water's uses outside the river channel, while the most pressing issues of river and water management in the Amazon are linked to in-stream uses. Consequently the countries of the Andean Amazon generally lack legal frameworks and institutional capacities to effectively manage water in its river channels for the benefit of in-stream uses.

That said, for much of the rural Andean Amazon, questions of national policies are largely irrelevant to the day-to-day realities of natural resource management. While some forms of resource use (e.g., petroleum extraction) are strongly influenced by national policies, extraction and use of local resources (including water, wood, land, and wildlife) by local people happen somewhat independently of national policies. Instead local economies, authorities, institutions, and cultural practices exert the greatest controls over resource management and extraction. At the local level, effective resource management is closely linked to education and land tenure—land owners and indigenous groups with tenure over their lands are confident that they will reap the rewards of good management and are therefore willing to make the effort and investment to manage more wisely. Rivers and their resources, on the other hand, are property of the state and therefore shared resources. Consequently river resources are subject to the irresponsible use and management that plague common resources. Effective management of common resources generally falls to government. But the national government presence is minimal in Amazon communities, and regional and local governments have very limited capabilities.

Outreach, Networking, and Capacity Building: Approach and Lessons Learned

From our planning stages, we worked to link AARAM activities to decision making and water resources management in Colombia, Ecuador, Peru, and Bolivia, but our approach evolved and changed over the course of the project. In the proposal stage, we signed letters of agreement with the government agencies responsible for monitoring water resources in each country (Table 19.1). Our goal was to ensure that AARAM activities built upon, and our data fed directly into, the national monitoring systems. We reasoned that the most effective way to build strong bridges between AARAM and those agencies was to bring agency staff into AARAM as graduate students. This technique proved effective, and over the course of the project, we trained one hydrologist from the National Meteorology and Hydrology Service of Peru (SENAMHI-Peru) (who has since been promoted to head of hydrology) and two hydrologists from the National Institute for Meteorology and Hydrology of Ecuador (INAMHI) (one of whom is now head of research in the institution).

Our experiences were, however, quite inconsistent among the agencies, and the key to long-term engagement was subcontracting the agency to conduct monitoring at existing stations within their networks. In each of the agencies, we found interest in adding gauging stations to their networks, but the construction and staffing of these stations was more expensive than our budget would allow. Only INAMHI remained engaged in the

Table 19.1. AARAM government partners at the initiation of the project

Country	Agency
Colombia	Instituto de Hidrología, Meteorología y Estudios Ambientales (IDEAM) (Institute for Hydrology, Meteorology and Environmental Studies)
Ecuador	Instituto Nacional de Meteorología e Hidrología (INAMHI) (National Institute for Meteorology and Hydrology)
Peru	Servicio Nacional de Hidrología y Meteorología (SENAMHI) (National Meteorology and Hydrology Service)
Bolivia	Servicio Nacional de Hidrología y Meteorología (SENAMHI) (National Meteorology and Hydrology Service)

project over the entire five years, largely because we subcontracted them to sample three of their existing stations and worked closely with them to model daily precipitation runoff processes in the upper Río Quijos basin (Galárraga and Torres 2001; Galárraga et al. 2004; Galárraga et al. 2005). Colombia's Institute for Hydrology, Meteorology and Environmental Studies (IDEAM) participated directly in the project for the first three years, as we subcontracted their hydrologists to participate in field campaigns sampling lowland rivers where no permanent stations were installed. SENAMHI-Bolivia dropped out of the project early because Bolivia had not yet ratified the IAI agreement, and funds could not be expended there. SENAMHI-Peru was engaged only through the involvement of their staff member studying within AARAM. This staff member did, however, develop a very useful model for predicting flows in the Rio Pachitea basin, and it is our understanding that this model has now been adapted to other basins in Peru (Ordoñez 2001).

In the first year of AARAM, we also made concerted efforts to inform and engage government agencies at the national level. We presented AARAM in meetings and seminars with a number of national agencies, including the Ministry of Environment of Colombia (in the presence of the vice minister), Instituto para el Ecodesarrollo de la Región Amazónica Ecuatoriana (ECORAE), Fundación para la Ciencia y la Tecnología (FUNDACYT) of Ecuador, Consejo Nacional de Ciencia y Tecnología (CONCYTEC) of Peru, and Instituto Nacional de Recursos Naturales (INRENA) of Peru. These meetings and seminars served to spread the word about AARAM and the IAI, but they did not result in any long-term collaborations. They also made clear to us that these agencies, and their modest resources, were committed to their own agendas and activities. There was interest on their parts to have AARAM conduct studies in support of their planned and ongoing activities, but they did not see a clear way to apply planned AARAM results to their needs. This proved to be a revelation about the divide between investigator-driven research priorities and the practical interests of policy makers.

Approaching national government partners at this stage of AARAM, even though still an early stage, was too late.

During the second and third years of the project, as national level partnerships failed to develop and as we began to understand better the dynamics of river resource management in the rural portions of the region, our emphasis shifted to the local level. Our primary strategy was to build partnerships with local governments and non-governmental organizations (NGOs) in the immediate vicinity of our field activities. Our strongest partnership with local authorities developed in the town of Oxapampa, Peru, where we established a semipermanent presence with co-funding from the Andrew W. Mellon Foundation. The local Oxapampa government took a real interest in our work and, as a consequence, began a community program designed to reduce the dumping of trash in the river, which was previously the preferred technique of waste disposal. Informative signs and trash cans were placed along the river near the town center. We also began an environmental education program in the local elementary schools, and this developed into an annual Día del Río (Day of the River) event in which local school children hosted a mini fair in the town square, presenting drawings, models, and posters of their studies on river processes and good management practices. In two cases, individual students in AARAM took the initiative to partner closely with the small communities in which they worked. In both cases, they produced educational materials and held community workshops presenting project results and discussing their implications.

Our most productive local partnerships were forged with development NGOs, whose staffs proved to be excellent complements to our technical teams. NGOs often make long-term commitments to working with individual communities, and they therefore enjoy a level of community trust that a three-to-five-year research program could not hope to achieve. They understand community needs and are able to see linkages between technical studies and community development that our team members would not. The ideal NGO partner is linked to larger international actors and activities, such that they understand better the context of the research themes emphasized by the IAI. For example, our most effective NGO partners were ProNaturaleza (of Peru), which is closely associated with the Nature Conservancy; the Instituto del Bien Común (of Peru), which has close ties to the Woods Hole Research Center; and EcoCiencia (of Ecuador), which has acted as the local implementer of several international initiatives. Here, too, the strategy of building bridges through students was effective. Two staff members of EcoCiencia completed graduate degrees with AARAM.

Future Challenges

The important lessons that emerged from the AARAM project will continue to challenge efforts to forge effective partnerships with policy makers and local stakeholders in the lesser-developed countries of the Americas. A first challenge is to adapt the science to the immediate and existing interests of government agencies. We learned that research activities that do not address immediate national interests, regardless of merit, are unlikely to stimulate collaboration. The second challenge is to understand the scale at which decisions about resource management should be made if one wishes to maximize the impact of science guidance. As we learned, for field-based research programs, it may be more effective to shift focus to policy makers and stakeholders at the local scale. In this case, it is wise to develop partnerships with local NGOs who can better link research products to the needs and development programs of communities.

Literature Cited

Ayllon, R. 2002. *Social and ecological evaluation of the use and management of fish resources in the Pachitea River Basin, Peruvian Amazon.* MSc thesis, Florida International University, Miami.

Blanco, A. 2005. *The impact of solid and liquid wastes from rural communities on the Chorobamba River, Oxapampa, Peruvian Amazon.* MSc thesis, Florida International University, Miami.

Celi, J. 2005. *The vulnerability of aquatic ecosystems of the (upper) Napo River basin to anthropogenic (activities) disturbances, Ecuadorian Amazon.* MSc thesis, Florida International University, Miami.

CONAM 2002. *Estrategia Nacional de Cambio Climatico.* Lima, Peru: Comision Nacional de Cambio Climatico (CONAM). (www.conam.gob.pe)

Cossio, E. 2002. *Socioeconomic controls on land use in riparian zones of the Pachitea Basin, central Andean Amazon, Peru.* MSc thesis, Florida International University, Miami.

Galárraga, R., M. Delgado, and M. F. Arias. 2001. Diagnóstico de las condiciones actuales de uso del agua en la cuenca del río Napo (Amazonía Ecuatoriana), dentro de un esfuerzo regional andino. Caso puntual dimensiones humanas-cuenca del río Napo-Amazonía Ecuatoriana. *Seminario Taller Internacional: Un enfoque integrado para la gestión sustentable del agua—Experiencias de cooperación.* Buenos Aires, Argentina.

Galárraga R., J. Molina, and M. Villacís. 2004. *Modelación matemática del proceso precipitación escurrimiento en la cuenca del río Quijos, cuenca alta del río Napo, Ecuador.* Final project report. Inter-American Institute for Global Change Research (IAI).

Galárraga R., and M. C. Torres. 2001. Water quality in the Napo River Basin (Ecuadorian Andean Amazonia). Andean Amazon Rivers Analysis and Management Project (AARAM). *Mountain Research and Development* 21(3):295–6.

Galárraga R., M. Villacís, M. Ayabaca, and J. Molina. 2005. *Daily precipitation-runoff modelling in an Andean Amazon river basin. Case study: The Quijos river basin in the upper Amazon Napo river basin in Ecuador.* San José, Costa Rica: Congreso V Foro Centroamericano y del Caribe de cuencas hidrográficas "desarrollo de cuencas hidrográficas."

McClain, M. E. 2002. The application of ecohydrological principles for better water quality management in South American rivers and wetlands. In McClain, M. E., ed. *The ecohydrology of South American rivers and wetlands.* Special Publication no. 6 of the International Association of Hydrological Sciences. (pp. 193–209)

McClain, M. E., L. M. Aparicio, and C. A. Llerena. 2001. Water use and protection in rural communities of the Peruvian Amazon. *Water International* 26:400–10.

McClain, M. E., and R. E. Cossio. 2003. The use and conservation of riparian zones in the rural Peruvian Amazon. *Environmental Conservation* 30:242–8.

McClain, M. E., and C. A. Llerena. 1998. El Manejo de Cuencas en la Selva: De los Andes a la Amazonia. *Agua y Riego* 11:11–3.

Mena, C.A., R. Bilsborrow, and M. E. McClain. 2006. Socioeconomic drivers of deforestation in the Napo River basin of Ecuador. *Environmental Management* 37:802–15.

Mittermeier, R. A., P. Robles Gil, and C. G. Mittermeier, eds. 1997. *Megadiversity: Earth's biologically wealthiest nations.* Monterrey, Mexico: CEMEX.

Ordoñez, J. 2001. *Analisis Hidrometeorologico y Aplicacion del Modelo de Simulacion IPH-MEN en la Cuena del Pachitea.* MSc thesis, Universidad Nacional Agraria La Molina, Lima, Peru.

Puentes, A. 2004. *Environmental conservation through solid waste management from the cradle to the grave in the Amazon: A case study in Puerto Bermudez, Peru.* MSc thesis, University of Florida, Gainesville.

20

Mediated Modeling for Integrating Science and Stakeholders: Impacts of Enhanced Ultraviolet-B Radiation on Ecosystem Services

Marjan van den Belt, Robert Costanza, Serge Demers, Susana Diaz, Gustavo A. Ferreyra, Sônia M. F. Gianesella, Evamaria W. Koch, Fernando R. Momo, and Maria Vernet

Mathematical modeling has historically been the province of experts (van den Belt 2004). When applied to policy questions, which span several disciplines and involve nontechnical stakeholders, the expert modeling approach is severely limited. Models for policy purposes usually become quite complex, and it is difficult to communicate results (or the underlying logic of the models) to decision makers in ways that are "credible" (Costanza and Ruth 1998). Models derive their credibility from two distinct sources: 1) the technical ability of the model to reproduce measured observations, and 2) the degree of "buy-in" to the assumptions of the model by stakeholders involved. Within a scientific discipline, the second source is often taken for granted. But in policy-relevant modeling, it often requires special attention. To achieve the necessary buy-in among the diverse stakeholders affected by a policy problem requires that the stakeholders participate in the modeling process. This chapter focuses on this participatory approach to model building in the context of the impacts of enhanced ultraviolet-B radiation (UVBR) on ecosystem services (Table 20.1). Enhanced UVBR is known from small-scale experiments to have several effects on organisms and ecosystems (Smith et al. 1992). But how those effects scale up and affect ecosystem services is not known. This chapter reports on both a mediated model-building process and the resulting model built jointly by scientists and stakeholders.

Table 20.1. The generic mediated modeling approach (as described in van den Belt 2004) with the three suggested stages with subactivities

1. Preparation	1.1. Identify stakeholders
	1.2. Set participant group
	1.3 Conduct introductory interviews
	1.4 Prepare a preliminary model
2. Workshops	2.1 Introduction
	2.2 Problem definition
	2.3 Qualitative model building
	2.4 Quantitative model building
	2.5 Simulation
3. Follow-up	3.1 Tutorial
	3.2 Evaluation

Methods

Mediated modeling is a process of facilitated model building in which stakeholders are involved in all aspects and steps (van den Belt 2004). Stakeholders decide what goes into the model they collaboratively construct, and therefore they understand the value (and limitations) of the results. In this project, the 15 participants included scientists involved in UVBR research and local stakeholders, both working on science and social issues.

Socioeconomic impacts of UVBR are difficult to pinpoint because of the lack of established cause-and-effect relationships at the local scale. The nature of the study prohibits an experimental design, and control groups are hard to locate. Historical observations provide some insight but have to be related to other trends.

Preparation

Much research in this program was on latitudinal differences in UVBR effects, and several different locations in the Americas were used for experiments. We selected Ushuaia, Argentina, as a location to make a connection with local stakeholders because it is an area with large UVBR fluctuations, with the associated social concerns and potential economic impacts—it is where the UVBR problem is most severe. The model constructed at Ushuaia was later scaled up to cover the area from the Arctic to the Antarctic (van den Belt et al. 2006).

The process of model construction is presented here. More detailed model results and scenarios can be found in van den Belt et al. (2006). Data available at this point for calibrating and testing the model are extremely scarce, so many results are speculative. The model has strong credibility due to stakeholder buy-in, but the technical calibration needs additional work.

First Workshop

The first workshop was held in November 2000 in Ushuaia, Argentina, and lasted four full days. A representative from the Centro Austral de Investigaciones Científicas (CADIC), the host institution, was instrumental in identifying the local stakeholders, based on previous knowledge of interested parties in the region. UVBR has been under study in Ushuaia since 1988, soon after the discovery of the antarctic ozone hole. Ten local stakeholders attended the first meeting with representatives from tourism, health (an ophthalmologist, a dermatologist), education, local government, business (fisheries), and scientists (meteorologists, several CADIC technicians). All participants contributed to every aspect of the workshop (van den Belt et al. 2006).

At the start, the goal of the workshop was stated, the four building blocks of the software (STELLA, High Performance Systems, Inc.) were introduced, and the questions the model should answer were formulated. They were the following:

- What are the direct and indirect socioeconomic impacts of UVBR?
- Can we define a synergy or feedback mechanism between different sectors represented in the model?
- Does the effect of UVBR on human health and ecosystem health change with scale?
- How does the time lag between chlorofluorocarbon (CFC) release and its aftermath (recovery of the ozone layer and ecosystem responses and human health effects) manifest itself?
- What is the influence of existing policies on local and global levels? Are there solutions to problems?
- Does global warming need to be in the model?

A preliminary model had been prepared in advance, based mainly on background literature. The group decided that the preliminary model provided a reasonable starting point for further discussion. This gave the group a running start but made it harder for the group to gain ownership of the model. This manifested itself in the way participants concentrated on their respective specialties rather than focusing on the system as a whole.

The qualitative model structure developed over the course of almost three days was the basis for a conceptual discussion. The group pushed the model into the unknown and generated future research questions based on it (Belzile et al. 2006; Costa et al. 2006; Diaz et al. 2006; Momo et al. 2006; van den Belt et al. 2006).

During the last day, all available data and information were worked into the model. In addition, the requirements for data gathering for the next two years were developed.

The workshop was closed with a reevaluation of the original problem definition and "the questions the model should answer."

The time between the first and the second meeting was dedicated to data gathering from scientists and stakeholders. In December 2001, Argentina's economy and research funding hit a low point, and data gathering became a low priority for our volunteer local stakeholders. Some participants changed jobs or no longer responded.

Second Workshop

In March 2004, a second stakeholder meeting took place in Ushuaia. The model was shown, feedback was provided, some scenarios were run, and the questions addressed by the model were revisited. Due to the national economy, which had temporarily pulled interests away from UVBR issues toward more acute livelihood concerns, there was less participation of the local stakeholders. It also became clear that the doom scenario for tourism, anticipated at the first meeting, had not happened.

Model Construction

The original intent was for the mediated modeling to function as a bridge between local stakeholders and investigators, to extend the research to the general public and attempt to make it usable for policy making. The model was built with STELLA, which uses four basic iconic building blocks: stocks, flows, auxiliary variables, and connectors. These can easily be identified in the model structures, respectively, as rectangles, valves, circles, and arrows.

The conceptual model structure that resulted from the first meeting was relatively well accepted by the group and encompassed much good thinking and discussion. Gathering the data turned out to be much more difficult than originally anticipated. The final version of the model was therefore simplified.

Two components of the model are presented here. Ecosystem services (Figure 20.1) are a constructive way to link different ecosystems and their contributions to human welfare (Boumans et al. 2002). This sector allowed us to sum up all the impacts of UVBR for the marsh and marine systems and express these impacts in monetary terms. While changes in the marine sector due to anthropogenic influences may *affect* global climate, marshes are expected to primarily be *affected by* climate change. Marshes cover relatively small areas when compared with oceans but have a high ecosystem service value per hectare.

Figure 20.2 shows the socioeconomic sector, using Tierra del Fuego as an example. Acute and chronic impacts of UVBR are translated into possible lost labor. The impacts on the tourist sector were not explored in great detail after the first meeting, since tourism was flourishing in 2004. The main conclusion was that it was desirable to prevent or mitigate UVBR impact through education.

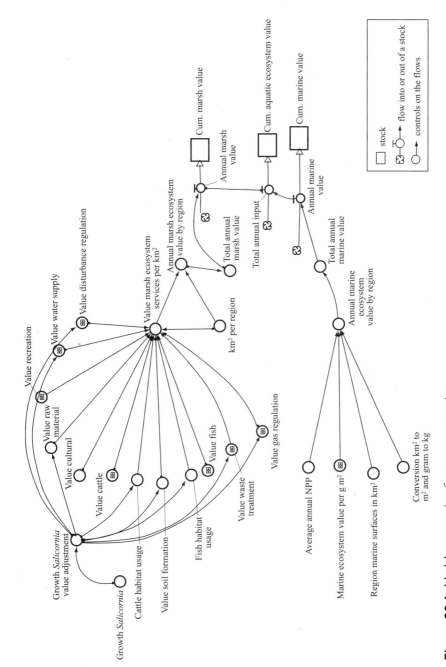

Figure 20.1. Model construction for ecosystem services

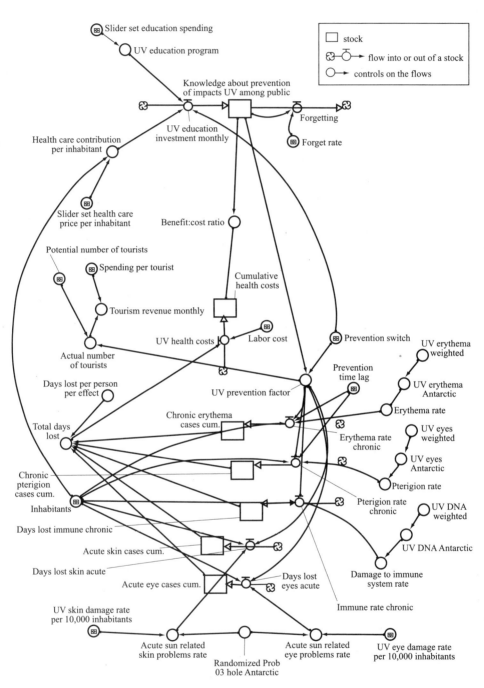

Figure 20.2. Model construction for socioeconomic factors

Conclusions

The model reflects the combined scientific and social elements identified as impacting life quality in regions affected by UVBR. It is therefore biased toward the world view of Ushuaia's population. Model construction and evaluation by scientists and policy makers brought agreement on the need to strengthen the links between research, education, and policy making.

Mediated modeling proved to be an effective tool to involve scientists and stakeholders in assessing UVBR impact on ecosystem services. This is a needed first step if UVBR research is to be effective in policy making. Mediated modeling facilitated stakeholder participation through problem formulation, data collection, and evaluation of the impact on society of problem mitigation. Ancillary benefits of the mediated modeling in Ushuaia were the education of local stakeholders on the UVBR effects and the subsequent dissemination of information at the community level (schools, radio, and government agencies).

The present model could be improved to include the effects of UVBR on freshwater and terrestrial ecosystem services, and it could include a greater number of ecosystem services. The improved model could be applied at different locations to provide effective solutions to the public.

Acknowledgments

This project would not have happened without local participants in Ushuaia who volunteered their time generously, including Miguel Isla, Osvaldo Barturen, Gabriel Karamanian, Sergio Luppo, Nemesio San Román, Oscar Bianciotto, Mónica Couly, Stella Domínguez, Sonia Pérez, Gloria Pereda, Graciela Ramacciotti, Marcelo Dragan, Ana Gamboa, and Luis Pinedo.

Literature Cited

Belzile, C., S. Demers, G. A. Ferreyra, I. Schloss, C. Nozais, K. Lacoste, B. Mostajir, S. Roy, M. Gosselin, E. Pelletier, S.M.F. Gianesella, and M. Vernet. 2006. UV effects on marine planktonic food webs: A synthesis of results from mesocosm studies. *Photochemistry and Photobiology* 82:850–6.

Boumans, R., R. Costanza, J. Farley, M. A. Wilson, R. Portela, J. Rotmans, F. Villa, and M. Grasso. 2002. Modeling the dynamics of the integrated Earth system and the value of global ecosystem services using the GUMBO model. *Ecological Economics* 41:529–60.

Costa, C. S. B., R. Armstrong, Y. Detrés, E. W. Koch, M. Bertiller, A. Beeskow, L. S. Neves, G. M. Tourn, O. A. Bianciotto, L. B. Pinedo, A. Y. Blessio, and N. San Roman. 2006. Effect of UV-B radiation on salt marsh vegetation: Trends of the genus *Salicornia* along the Americas. *Photochemistry and Photobiology* 82:878–86.

Costanza, R., and M. Ruth. 1998. Using dynamic modeling to scope environmental problems and build consensus. *Environmental Management* 22:185–95.

Costanza, R., F. Andrade, P. Antunes, M. van den Belt, D. Boersma, D. Boesch, F. Catarino, S. Hanna, K. Limburg, B. Low, M. Molitor, J. Pereira, S. Rayner, R. Santos, J. Wilson, and M. Young. 1998. Principles for sustainable governance of the oceans. *Science* 281:198–9.

Diaz, S., C. Camilion, G. Deferrari, H. Fuenzalida, R. Armstrong, C. Booth, A. Paladini, S. Cabrera, C. Casiccia, C. Lovengreen, J. Pedroni, A. Rosales, H. Zagarese, and M. Vernet. 2006. Ozone and UV radiation over southern South America: Climatology and anomalies. *Photochemistry and Photobiology* 82:834–43.

Momo, F., E. Ferrero, M. Eöry, M. Esusy, J. Iribarren, G. Ferreyra, I. Schloss, B. Mostajir, and S. Demers. 2006. The whole is more than the sum of its parts: Modelling community level effects of UVR in marine ecosystems. *Photochemistry and Photobiology* 82:903–8.

Smith, R. C., B. B. Prézelin, K. S. Baker, R. R. Bidigare, N. P. Boucher, T. Coley, D. Karentz, S. McIntyr, H. A. Matlick, D. Menzies, M. Ondrusek, Z. Wan, and K. J. Waters. 1992. Ozone depletion: Ultraviolet radiation and phytoplankton biology in Antarctic waters. *Science* 255:952–8.

van den Belt, M. 2004. Mediated modeling: A system dynamics approach to environmental consensus building. Washington, D.C.: Island Press.

van den Belt, M., O. A. Banciotto, R. Costanza, S. Demers, S. Diaz, G. A. Ferreyra, E. W. Koch, F. R. Momo, and M. Vernet. 2006. Mediated modeling of the impacts of enhanced UV-B radiation on ecosystem services. *Photochemistry and Photobiology* 82:865–77.

21

ENSO and Risk Management: Natural and Social Sciences, Policy Implications, and Stakeholder Participation

Allan Lavell

The Research Theme: Problems and Relevance

The considerations presented in this chapter derive from the research, training, and educational objectives and strategy guiding the ENSO Risk and Risk Management Project undertaken between 2000 and 2005 by the Latin American Network for the Social Study of Disaster Prevention (through its associate institutions and researchers) in eight continental American countries, all within the framework of the Inter-American Institute for Global Change Research (IAI) Collaborative Research Network (CRN) process.

The project design was guided by two needs:

- to understand and dimension changes in the hazard, vulnerability, and risk patterns associated with the El Niño/Southern Oscillation (ENSO) phenomenon, and climate variability in general, over the period 1970 to 2003 (with *risk* defined as the probability of future social and economic loss and damage resulting from a dynamic interaction between hazard intensity, magnitude, and exposure and existing human vulnerability conditions)
- to analyze used approaches to disaster and risk management and propose changes to these in accordance with the scientific research results of the project, providing a training and educational basis for promoting such changes

The project's central concern was to generate a territorially disaggregated disaster database in order to provide information with which to analyze the changing spatial, social, and semantic patterns of disaster risk, loss, and damage associated with ENSO- and non–ENSO-generated hydrometeorological hazards. DESINVENTAR is a database that allows the registering at the local or municipal level of all hazard-related events associated with any level of reported economic and social loss (www.desenredando.org). This information and the ensuing analysis offer a basis for the publication and dissemination of scientific results (including the development of an interactive regional ENSO documentation system open to all interested parties), and the development of training and educational modules on ENSO risk and risk management, aimed at different social actors, including direct and indirect stakeholders and decision makers.

The project's relevance may be understood when it is noted that ENSO disaster and risk management to date has concentrated predominantly on anticipating hazard events and attempting to provide preparedness and response mechanisms that allow a short-term decrease in disaster loss. Little has been done in the framework of ENSO and its relationship to more short-term, recurring aspects of climate variability in order to provide a basis for risk reduction and risk and disaster prevention and mitigation work. Few disaster agencies go beyond the response and rehabilitation aspects. Training and educational modules dealing with the complexities of ENSO risk management are not readily available to date.

This brief description of project objectives immediately reveals the policy nature of research, its pertinence to diverse stakeholders, and the necessary incorporation of natural and social science contributions in fostering an understanding of processes and intervention procedures. Projects are conceived around topics and objectives, concepts, hypotheses, and method. The nature of the natural science–social science interaction, the policy-making influences, and types of stakeholder participation sought or achieved will and should depend on project conception and project goals. In principle there can be no absolute or single correct position or approach. What is important is that these central aspects be given due consideration, inform the work undertaken, and be based on an explicit framework for contributing to science-based policy impacts with active stakeholder involvement. These aspects are among the central concerns of this chapter.

Key Research Findings

ENSO risk and overall risk associated with climate variability is essentially determined by the characteristics of prevailing development models and their consequences in terms of human exposure and vulnerability and in terms of environmental degradation and change. No standard or regular pattern of hazard impact exists at the regional or local levels, or between countries over successive ENSO periods. Inter-ENSO variability in risk conditions and levels is common. Consequently, even under similar hazard conditions, loss and damage vary greatly between periods. Accentuated human vulnerability and deteriorating environmental conditions guarantee increases in risk even when hazard intensity or magnitude is relatively constant. Risk can even increase under decreasing hazard conditions because of increased exposure and vulnerability of population groups.

The unpredictable nature of many lower-scale territorial level impacts (subregional, local) is as yet an insurmountable challenge for short-, medium-, and long-term policy formulation. Despite this uncertainty, response to new events is generally based on the notion of continuity and repetition of impacts. Many times this assumption results in erroneous policy decisions. ENSO-hazard-related events affecting any one region or zone are often similar to the hydrometeorological hazards generated by climate variability in non-ENSO years. Risk is associated with both. And, often, this risk may be greater in non-ENSO than in ENSO years. ENSO is, therefore, only one (if an important) component of overall climate variability and risk in many regions and zones (the tropical humid zones of Central and South America, for example). In others it is the principal, if not only, factor generating "abnormal" hazard conditions (for example, in the north of Peru and in parts of Ecuador and Bolivia). The only realistic way to deal with ENSO hazards is, therefore, to see them as part of a complex succession of events associated with overall climate variability and probably influenced by global climate change. Society must then promote holistic and integral, cross-temporal and cross-hazard approaches to management. These must be site specific and involve the full range of national and, particularly, local stakeholders.

Outreach Aspects: Successes, Failures, and Surprises

The study of ENSO and climate variability and its relations to risk, damage, and loss has shown a clear bias toward the physical/geophysical, climatological, or oceanological aspects and the definition of *hazard conditions*. Although not always explicit, there is a tendency to unilaterally associate hazard with risk and, thus, with disaster. Society is seen to be a victim but not a contributing cause. The perceived solution to this is to increase our ability to predict events, provide early warning, and thus reduce risk, loss, and damage by providing people with the means to protect themselves and their livelihoods in a reactive, short-term fashion.

Understanding ENSO-related disaster risk requires an analysis of the territoriality and temporality of hazard incidence and the semantics of *hazard*, among other factors. Without hazards there are clearly no disaster risks. However, in saying this, it cannot be assumed that risk and disaster are unilaterally "caused" by hazard itself. For risk to exist, vulnerable populations and economies must also exist, and this vulnerability (as well as many hazards) is socially constructed—the result of social processes. These processes may lead to the location of vulnerable groups in exposed areas, to the generation of new hazards through environmental degradation, or to increasingly vulnerable human populations.

The present study and its training and educational activities challenge the "physicalist" (see Hewitt 1983), hazard-based paradigm used in the explanation of risk and disaster. Throughout the project, the DESINVENTAR disaster database was used as a source for the analysis of damaging events and their causal processes. The semantic, temporal, social, and territorial changes in hazard incidence have permitted a debunking of the hazard paradigm and the provision of evidence for the social nature of risk construction. Moreover, the conclusions trace paths toward reforms of public policy and to the importance of new educational and training paradigms, with increased stakeholder participation.

Although project results have allowed a broad confirmation of the above-mentioned points, the DESINVENTAR database is as yet limited in many ways, and far more work and constant updating are required in order to ratify and expand the information available. The lack of systematic collection of disaster information at local scales, along with a preponderance of indirect sources, such as newspapers, inevitably compromises the precision of the information. Moreover, there is a serious lack of available information on many of the economic loss and damage categories so necessary for adequate dimension of risk and disaster.

Project training and educational activities have permitted a pilot approach to policy implications and stakeholder involvement. The training and educational materials developed with the project are among the only ones available from a social risk perspective. However, a more generalized process of dissemination is required, and a more comprehensive inclusion of the varying stakeholders must be sought. These range from local government and nongovernmental actors through to international agencies. At present ENSO, climate variability, and climate change topics are still subject to professional specialization instead of being treated as a single integrated problem with commensurate policy and stakeholder needs. One welcome aspect of the project and its results is that work with local actors shows that they are far more likely to comprehend and follow more integral risk analysis and management approaches than national or international actors would. Specialization tends to increase the higher up the territorial and social scale one goes.

Networking: The Social and Natural Science Interface and Reaching Out to Stakeholders

Knowledge of risk and disaster can be achieved only through the dynamic and interactive development of the natural and social sciences. The nature of this relationship and its hierarchical aspects are determined by the object of study. In the present case, risk and risk management are conceived as socially constructed realities and activities while natural sciences provide substantive information and process inputs on relevant physical events and hazard contexts. However, both risk and risk intervention are socially constructed and construed, such that hazard is not the dominant concern when it comes to the explanation of risk. Nor is hazard the principal issue involved in risk management. Hazards are given social meaning through the application of sound and tried social science concepts and methods. There can be no mechanistic relationship between natural science findings and public or private policy formulations, given that risk is socially constructed and many times construed in socially subjective and relative ways. Risk is thus subject to interpretations, decisions, and controls that are defined in social and in socially subjective scientific manners. Information and interpretations with regard to ENSO and other manifestations of climate variability must necessarily be socially contextualized, actualized, filtered, and elaborated.

Direct stakeholder participation was not attempted in the project, given that the research objectives included a dimensioning of the different ways in which stakeholder interests dictate risk construction processes and guide intervention schemes and strategies. Stakeholders being among the "objects" of study are, therefore, not an appropriate component of the research process as such, but rather a market and participant in research results and debate. Despite this lack of direct involvement, the stakeholders are integrated into the project through its debating forum and project dissemination strategy. The project created parallel structures and options for participation. These took the form of stakeholder and policy networks, created or strengthened, in general, through holding in-country research conferences and seminars.

Capacity Building: From Research to Policy Making

The ENSO risk project aimed to provide better explanations of risk and risk management procedures, integrating in a holistic fashion findings from the social and natural sciences. This can be seen as better science. At the same time, the policy implications of this science were considered apart—a sequential, specialized area of concern. This does not, of course, preclude the inevitable coloring of social science research with social, political, and methodological biases. Natural science cannot afford these biases, and "purity" is always demanded by the public and politicians.

Research results have profound policy connotations. However, the translation of these into alternative policy frameworks was not a viable objective for research, but rather a co-related, sequenced objective to be achieved through other social and political mechanisms. We must often satisfy ourselves with making good science and then take on the policy implications and dissemination required. The search for good science does not mean that scientists in general cannot be advocates and movers of policy change outside the "laboratory." Policy making, when seen from a social and political science angle, may be informed by science but is decided by dominant stakeholder and group interests. The sociological analysis of science and decision has clearly demonstrated this over the years, for example, in the debates and conclusions about fluoride, vitamin C and cancer, stem cell research, and global climate change.

In the case of the ENSO risk project, while the overriding goals of finally influencing policy decisions were strongly present in project design and hypotheses, they were not incorporated methodologically into the basic project components of data collection and research. However, the data collection and analysis had implicit and explicit policy change goals, based on the appreciation of ongoing, flawed intervention practice. Once research is completed and results are published or distributed, one would clearly hope for direct and indirect policy impacts.

Challenges for the Future

Two principal types of challenge may be seen to exist with regard to risk associated with climate variability and climate change. First is the ongoing challenge of bringing natural and social science practitioners together in the design and application of research projects that start from common objectives, conceptual viewpoints, and interdisciplinary perspectives. Such projects aim at the reduction of societal risk from climate variability and hazard generated by climate change. Today, after many years of insistence on the need for common bases for research, the topic is still dominated by unilateral natural or social science perspectives. The introduction of a human dimension into natural science projects is usually achieved by the "adding on," but not the integrating, of a social component, while many social science projects only make a cursory mention of natural and hazard processes.

Second, an attempt must be made in terms of education and practice to break down the false division and professional bias implied in the separate, as opposed to integrated and holistic, treatment of ENSO, other climate variability hazard contexts, and the climate change problematic, with its adaptation and mitigation approaches. Risk management must be seen as a multi-hazard, multi-context informed, continuous activity designed within the framework of sustainable development objectives and goals. The false separation of ENSO, climate variability, and climate change analysis and management can lead only to diseconomies of scale and impact.

Literature Cited

Hewitt, K. 1983. The idea of calamity in a technocratic age. In Hewitt, K., ed. *Interpretations of calamity*. Essex: Longman Press.

Additional Reading on the Role of Risk

Bankoff, G. et al., eds. 2003. *Mapping vulnerability*. U K: Earthscan Publications. (http://www.earthscan.co.uk/)

Bora, A. 2005. *Stakeholder involvement, science policy and practical decision making: A critical review*. Paper presented at the PRIME Annual Conference, Manchester, January 8, 2005, session 2A.

Cannon, T. 2003. *Vulnerability analysis, livelihoods and disasters. Components and variables of vulnerability: Modelling and analysis for disaster risk management*. Paper presented at the Expert Meeting on Disaster Risk Conceptualization and Indicators Modelling, Technological University of Cataluña, July 9–11, 2003. Organized by IDEA, National

University of Colombia, Manizales, within the framework of the IADB Information and Indicators Programme for Risk Management.

Cahn, M. Forthcoming. *Linking science to decision making in environmental policy: Bridging the disciplinary gap.* Cambridge, Mass.: MIT Press.

Cardona, O. D. 1999. Environmental management and disaster prevention: Holistic risk assessment and management. In Ingleton, J., ed. *Natural disaster management, IDNDR.* London: Tudor Rose.

Cash, D., and S. Moser. 1998. *Information and decision making systems for the effective management of cross-scale environmental problems.* Paper presented at the workshop on Local Response to Global Change: Strategies of Information Transfer and Decision Making for Cross Scale Environmental Risks, Belfer Center for Science and International Affairs, John F. Kennedy School of Government, Harvard University, January 29–30, 1998.

Hewitt, K. 1997. *Regions of risk: A geographical introduction to disasters.* Essex: Longman.

Jacobs, K. 2003. *Connecting science, policy and decision making: A handbook for researchers and science agencies.* Boulder, Colo.: NOAA Office of Global Programs.

Lavell, A. 1993. Ciencias sociales y desastres naturales en América Latina: Un encuentro inconcluso. In Maskrey, A. *Los desastres no son naturales.* Editorial Tercer Mundo, Bogotá, Colombia.

Lavell, A. 1996. Degradación ambiental, riesgo y desastre urbano: problemas y conceptos. In Fernández, M. A. *Ciudades en riesgo.* Lima, Perú: LA RED, USAID.

Lavell, A. et al. 2003. *La gestión local del riesgo. Nociones en torno al concepto y la práctica.* CEPREDENAC- PNUD. (www.cepredenac.org)

Martin, B., and E. Richards. 1995. Scientific knowledge, controversy and public decision making. In Jasanoff, S. et al., eds. *Handbook of science and technology studies.* Newbury Park, Calif.: Sage. (pp. 506–26)

Snover, A., E. Miles, and A. Hamlet. 2003. *Learning from and adapting to climate variability in the Pacific North West.* In background papers in preparation for the Insights and Tools for Adaptation: Learning from Climate Variability Workshop, November 18–20, 2003, Washington, D.C. Washington, D.C.: NOAA Office of Global Programs. (pp. 167–79). (Available at http://www.cses.washington.edu/db/pubs/topic8.shtml)

Wilches Chaux, G. 1998. *Auge, caída y levantada de Felipe Pinillo, mecánico y soldador. Guía de La Red para la gestión local de los desastres.* La Red, Quito, Ecuador.

Wisner, B. et al. 2003. At risk: Natural hazards, people's vulnerability and disasters. 2nd ed. Oxford: Routledge Press.

22

Diagnostics and Prediction of Climate Variation and Human Health Impacts

Ulisses Confalonieri

The main objective of this project, involving Brazil, Venezuela, the United States, Jamaica, Mexico, and Colombia, was to generate regional information on the linkages between climatic factors and two major endemic infectious diseases: malaria and dengue fever. It was expected that this new knowledge would support public health programs by providing new tools for disease control. Although the disciplinary background of the investigators had a strong bias toward the natural sciences (epidemiology, entomology, and climatology) it had an important "human dimensions" component, since it dealt with human diseases and their determinants, not just with biological hazards. This chapter is based on the research process that took place during the Collaborative Research Network (CRN) project and on the experience of the coinvestigators in trying to develop applications for the research products with governmental agencies.

Framing the Science

The different scientific disciplines were used in the participating countries to varying degrees, depending on the capabilities of participating institutions and the partnerships developed. Since malaria and dengue are transmitted by different mosquito vectors, entomologists played an important role in the research process. A great deal of the modulating influence of climate on the dynamics of these diseases is mediated by the effects of the physical factors on the life cycle of the vectors.

The collection of socioeconomic data in some of the participating countries was part of the epidemiological studies. Even though climate-disease linkages were the primary focus of the research, infectious diseases are determined by a host of factors, many of them non-climatic. Therefore, the need to assess "confounding factors" was recognized early in the process.

Scientific achievements of the project included the following:

- documentation of regional associations between patterns of climate variability and the incidence of dengue fever in Mexico, Jamaica, and Brazil
- demonstration of the effects of the El Niño/Southern Oscillation (ENSO) phenomenon on the regional dynamics of malaria, especially in Colombia
- development of new methodological approaches for the analysis of climate-health linkages in long time series of data
- understanding of the role of local climate variation in the incidence of malaria in Brazil and Venezuela
- and the demonstration of the role of microclimatic factors in the population dynamics of mosquito vectors in the tropical rain forest

Networking and Outreach

The project was able to make an important contribution to regional networking in the climate-health arena. Research and training capacity in institutions of developing countries was established with the creation of "reference groups," and climate was included in the public health research agendas of these countries. The project facilitated the involvement of investigators in national and international fora and scientific assessments related to the health impacts of climate and other environmental changes, such as the Intergovernmental Panel on Climate Change (IPCC), the Millennium Ecosystem Assessment, and the Earth System Science Partnership. The research project utilized both secondary and primary epidemiological and climatological data, and it relied greatly on data generated routinely by governmental agencies. It developed new applications of weather and climate forecasts to the prediction of disease occurrence in developing countries. The US cooperators were involved either in outreach activities or as climatological advisors. Cooperators from Brazil, Mexico, Jamaica, Venezuela, and Colombia compiled secondary data, generated new data, and provided data analysis. In these countries, there was a continued involvement of the investigators with local or regional health departments in charge of epidemiological surveillance, as providers of data.

Stakeholder participation was expected during production and compilation of data and in the use of the products developed from the analysis of the climate-health linkages. These products support the development of "early warning systems" for the diseases. Therefore the expected users were the public health officers involved in disease control activities, although the models constructed did not become fully operational for disease prediction during the course of the project.

Another community that can be considered a stakeholder group are climatologists not directly involved in the project. Some field entomological research was associated with experiments in climatology, for instance, in Brazil at one of the meteorological towers of the international Large-Scale Biosphere-Atmosphere Experiment in Amazonia (LBA). The CRN has become associated with this large-scale project and allowed a new biological application for the vast array of physical data generated by the climatological experiment.

On a different level, stakeholder involvement also took place with the Pan American Health Organization (PAHO). Participation in the organization of a climate-health workshop in Barbados in 2002 was important in bringing the climate-health issue to a broader audience. A similar event took place in Jamaica in November 2005, through an Inter-American Institute for Global Change Research (IAI) Training Institute on Climate and Health. An important outreach activity and product, developed early in the project, was a Web site that improved the communication of the researchers with the policy community.

The likelihood that the scientific products of this project will have an influence on policy decisions at national or international levels depends on the demonstration that climate is a key determinant of the dynamics of the diseases and on the development of practical tools for disease monitoring and prediction in relation to climate variability.

To varying degrees, all participating countries were able to demonstrate relevant climate-health linkages at local, regional, and national levels. However, integration with health officers in their governmental agencies was extremely "variable in time." This caused difficulties in incorporating the research results into epidemiological surveillance and disease control programs, largely because of the existence of other "operational priorities" in the health agencies. Disease control officers were keen to put efforts and resources into well-known activities, such as spraying for the control of mosquito populations, or to treat diseased individuals rather than to devote time to the development of new approaches such as the early warning systems. These systems are not meant to replace the traditional disease control interventions but rather to help epidemic preparedness and the targeting of scarce public resources. Demands on project personnel were more likely to occur when extreme weather events caused impacts on their countries, when outbreaks of climate-sensitive diseases occurred, or even when governments were requested to refer technical personnel to international meetings.

An important factor missing was the capacity to demonstrate the cost-effectiveness of the early warning systems. However, interest has been shown by the Director of Disease Prevention and Control in the Ministry of Health of Jamaica in a proposed simple climate-dengue early warning system, by the Ministry of Social Protection of Colombia in a geographic information system to be freely distributed as a standard platform for malaria databases, by the national directorate for vector control in Mexico in the incorporation of climatic indicators in epidemiological surveillance programs, and by the division of environmental health of the Brazilian Ministry of Health in climate-health issues in the semiarid region of the country. The project has called the attention of the health ministries and secretariats to the role of climate in the epidemiological dynamics of important endemic diseases. It has also produced useful tools for epidemiological surveillance and analysis, important for the early detection of the impacts of climate variability and change.

Capacity Building

The research has provided opportunities for graduate and undergraduate students in several areas (from geography to biology and veterinary medicine) to participate in research activities. Besides creating career opportunities for young scientists and students in developing countries, the project has supported capacity building of academic and governmental institutions. In Brazil, Mexico, and Venezuela, public health research and training institutes were able to address climatic drivers of infectious diseases in their countries. This was a considerable step forward in shaping the health research agenda in Latin America since the three participating institutions are part of the national ministries of health, which are in charge of the formulation of the health policies. The academic departments of climate and hydrology in Colombia and Jamaica were able to assemble multidisciplinary teams to investigate climate-health linkages. Another important aspect was the development of training courses for local technicians.

Future Challenges

Lessons learned from science-policy integration issues in this CRN project are the following:

- A good degree of integration of different scientific disciplines has been achieved in the research, but this has happened mostly under the umbrella of the natural sciences.
- Effective participation of the social sciences in the development of the scientific products and their applications has been only marginal.
- Health authorities in most involved countries have shown an interest in the impacts of climate. However, they seem overburdened with so many other pressing health care issues that a good assimilation of the scientific outputs of the project has not been possible so far.
- The dissemination of the scientific results of the project to decision makers outside the project's team has not yet happened.

There are many future challenges related to both the science base and the interface with public policies. From the science perspective, there is still a need to build longer and easily accessible time series of epidemiological data, especially at the local scale. Different analytical modes should be applied in order to clarify issues generated by findings at the regional scale that may not be in agreement with results from local studies.

From the policy perspective, the major challenge will be the demonstration that the accumulated knowledge on climate-disease linkages can be translated into cost-effective early warning systems, to be incorporated in the operational routine of health departments.

About the Editors

HOLM TIESSEN is Director of the Inter-American Institute for Global Change Research (IAI) and Professor on leave from the Georg-August Universität of Göttingen, Germany.

GERHARD BREULMANN, a terrestrial ecologist, is Science Officer at the IAI in São José dos Campos, Brazil.

MIKE BRKLACICH is Professor of Geography and Environmental Studies at Carleton University in Ottawa, Canada, and chairs the IAI's Scientific Advisory Committee.

RÔMULO S. C. MENEZES is Assistant Professor in the Department of Nuclear Energy at the Universidade Federal de Pernambuco, Recife, Brazil.

Contributors

Azócar, Aura
Facultad de Ciencias
Instituto de Ciencias Ambientales y
Ecologicas (ICAE)
Universidad de los Andes
Mérida 5101
Venezuela

Baethgen, Walter
Director, Latin America/Caribbean
Program
IRI Columbia University
Palisades, NY 10964-8000
USA

Ballester, Maria Victoria Ramos
CENA-USP
Av. Centenario 303
13416-000 Piracicaba, SP
Brazil

Beckers, Theo
Telos, Institute for Sustainable
Development
Tilburg University, P.O. Box 90153
5000 LE Tilburg
The Netherlands

Berbara, Ricardo Louro
UFRRJ, Departamento de Solos
BR 465 KM 7
23851-960 Seropédica, RJ
Brazil

Breulmann, Gerhard
IAI Directorate, c/o INPE
Av. dos Astronautas 1758
12227-010 São José dos Campos, SP
Brazil

Brklacich, Mike
Dept. of Geography & Environmental
Studies
Carleton University
1125 Colonel By Drive
Ottawa, K1S 5B6, Ontario
Canada

Brown, Irving Foster
WHRC-UFAC, Parque Zoóbotanico
Universidade Federal do Acre
69915-900 Rio Branco, Acre
Brazil

Caetano, Neto Ernesto dos Santos
Center for Atmospheric Sciences
National Autonomous University of
  Mexico
Av. Circuito Exterior s/n
Ciudad Universitaria
04510 Mexico City
Mexico

Câmara, Gilberto
Instituto Nacional de Pesquisas
  Espaciais INPE - DPI
Av. dos Astronautas 1758
12227-010 São José dos Campos, SP
Brazil

Campos, Edmo J. D.
Dept. of Physical, Chemical and
  Geological Oceanography,
  Oceanographic Institute
University of Sao Paulo
Pça. do Oceanográfico 191
05508-900 São Paulo, SP
Brazil

Castellanos, Alejandro
DICTUS, Universidad de Sonora
Apartado Postal 54
Hermosillo, Sonora
Mexico

Chazan, May
Dept. of Geography & Environmental
  Studies
Carleton University
1125 Colonel By Drive
Ottawa, K1S 5B6, Ontario
Canada

Confalonieri, Ulisses
PMAGS/ENSP FIOCRUZ
Av. Brasil, 4036/703 Manguinhos
Rio de Janeiro, RJ
Brazil

Costanza, Robert
Gund Institute for Ecological
  Economics
University of Vermont
Burlington, VT 05445
USA

Dawe, Andrew
Dept. of Geography & Environmental
  Studies
Carleton University
1125 Colonel By Drive
Ottawa, K1S 5B6, Ontario
Canada

Demers, Serge
Institut des Sciences de la Mer de
  Rimouski
310 Allée des Ursulines
Rimouski, Québec G5L 3A1
Canada

Diaz, Susana
CADIC/CONICET CC92
H. Yrigoyen s/n
9410 Ushuaia
Tierra del Fuego
Argentina

Ferreyra, Gustavo A.
Instituto Antártico Argentino
Cerrito 1248
1010 Buenos Aires
Argentina

Filmer, Paul E.
Program Director
US NSF, GEO/OAD & EAR/GEO
4201 Wilson Blvd, Suite 785
Arlington, VA 22230
USA

Galárraga-Sanchez, Remigio H.
Escuela Politecnica Nacional (EPN)
Departamento de Ciencias del Agua
  (DCA)
Ladron de Guevara E11-253
Apartado Postal 17-01-2759, Quito
Ecuador

Gallardo Klenner, Laura
Meteorología Química
Centro de Modelamiento Matemático
Universidad de Chile (UMI CNRS
  2807)
PC 170-3, Santiago
Chile

Garvin, Theresa
Director, Community, Health &
  Environment Research Centre,
  Human Geography
Department of Earth & Atmospheric
  Sciences
University of Alberta
Calgary, Alberta
Canada

Gianesella, Sônia M. F.
Instituto Oceanográfico da
Universidade de São Paulo
Pça do Oceanográfico 191
05508-900 São Paulo, SP
Brazil

Göbel, Barbara
Instituto Ibero-Americano
Potsdamer Str. 37
D-10785 Berlin
Germany

González, A. Juan
Fundacíon Miguel Lillo
Instituto de Ecología Vegetal
Miguel Lillo 251
Tucumán 4000
Argentina

Herpin, Uwe
University of São Paulo, NUPEGEL,
  ESALq CP 9
13418-900 Piracicaba, SP
Brazil

Herrera-Peraza, Ricardo
Instituto de Ecología y Sistemática
Ministerio de Ciencia, Tecnología y
  Medio Ambiente
La Habana
Cuba

Ingram, John
GECAFS Executive Officer / COST
  ESSEM DC Vice Chair
GECAFS International Project Office
Environmental Change Institute
Oxford University Centre for the
  Environment
South Parks Road, Oxford OX1 3QY
United Kingdom

Jiménez-Osornio, Juan J.
Universidad Autonoma de Yucatán
  PROTROPICO
Apartado Postal 28, Cordemex
Mérida, 97110
Mexico

Jutro, Peter
Deputy Director for Science & Policy
National Homeland Security Research
  Center
Office of Research & Development
U.S. Environmental Protection Agency
  (8801R)
1200 Pennsylvania Avenue, NW
Washington, DC 20460
USA

Klink, Carlos
Depto. de Ecologia, Universidade de
  Brasilia CP 04457, Campus
  Universitario Darcy Ribeiro - Asa
  Norte
70910-970 Brasilia, DF
Brazil

Koch, Evamaria W.
Horn Point Laboratory
University of Maryland
Center for Environmental Science
P.O. Box 775
Cambridge, MD 21613
USA

Krusche, Alex
CENA-USP
Av. Centenario 303
13416-000 Piricicaba, SP
Brazil

Lavell, Allan
Secretariat General, Latin American
  Social Sciences Faculty (FLACSO)
and
Latin American Network for the Social
  Study of Disaster Prevention (LA
  RED)
Del McDonalds de Curridabat
200 sur, 25 este, San José
Costa Rica

Leinen, Margaret S.
Chief Science Officer
Climos, Inc.
119 South Columbus Street
Alexandria, VA 22314
USA

Liu, Kam-biu
James J. Parsons Professor of
  Geography Department of
  Geography & Anthropology
Louisiana State University
227 Howe-Russell, Geoscience
  Complex
Baton Rouge, LA 70803-4105
USA

Llerena, Carlos A.
Universidad Nacional Agraria La
  Molina (UNALM)
Departamento de Manejo Forestal
  (DMF)
Facultad de Ciencias Forestales
Av. La Universidad La Molina s/n
P.O. Box 456
Lima 12
Peru

Luckman, Brian
Department of Geography
University of Western Ontario
London, N6A 5C2, Ontario
Canada

Magaña, Victor
Universidad Nacional Autónoma
  de México (UNAM)
Centro de Ciencias de la
  Atmósfera
Becal 510 Col Heroes de
  Padierna
Ciudad Universitaria
Mexico DF 04510
Mexico

McClain, Michael
Dept. of Environmental Studies
Florida International University
Miami, FL 33199
USA

Melillo, Jerry
The Ecosystems Center
Marine Biological Laboratory
7 MBL Street
Woods Hole, MA 02543-1301
USA

Menezes, Rômulo S.C.
Depto de Energia Nuclear
Universidade Federal de
  Pernambuco
Cidade Universitaria
50740-540 Recife, PE
Brazil

Momo, Fernando R.
Universidad Nacional de General
  Sarmiento
Instituto de Ciencias
J. M. Gutiérrez 1150
B1613GSX Los Polvorines
Argentina

Nobre, Carlos A.
Instituto Nacional de Pesquisas
  Espaciais (INPE)
Av. dos Astronautas 1758
12227-010 São José dos Campos, SP
Brazil

Noellemeyer, Elke
Secretaria de Ciencia y Técnica
Facultad de Agronomía
Universidad Nacional de La Pampa
Casilla de Correo 89
Santa Rosa, RA-6300, L.P.
Argentina

Nuñez, Mario
Universidad de Buenos Aires (UBA)
Depto de Ciencias de la Atmosfera y
  Oceanos
CIMA Pabellon II Piso 2
Ciudad Universitaria
Buenos Aires 1428
Argentina

Piola, Alberto
Servicio de Hidrografia Naval
  (Argentina)
Departamento de Oceanografia
Av. Montes de Oca 2124
Buenos Aires 1271
Argentina

Plocq-Fichelet, Véronique
SCOPE
51 bd de Montmorency
75016 Paris
France

Reyes-Knoche, Susanne
Institute for International Public Law
& European Law
Dept. of International Economic Law
& Environmental Law
University of Göttingen
Platz der Göttingen Sieben 5
D-37073 Göttingen
Germany

Ruiz, Jose Efraín
Facultad de Ciencias
Departamento de Química
Universidad de los Andes
Carrera 1 no. 18A 10 Bogota
Colombia

Salcedo, Ignácio H.
Depto de Energia Nuclear
Universidade Federal de Pernambuco
Cidade Universitaria
Recife PE 50740-540
Brazil

Sanchez-Azofeifa, Arturo
Earth and Atmospheric Sciences
Department
University of Alberta
Edmonton, T6G 2E3, Alberta
Canada

Silva, Juan F.
Universidad de los Andes (Venezuela)
(ULA)
Instituto de Ciencias Ambientales y
Ecologicas (ICAE)
Facultad de Ciencias, ULA
Merida 5101
Venezuela

Stauffer, Mark D.
President (retired), Potash &
Phosphate Institute of Canada
202-415 Heritage Crescent
Saskatoon, SK S7H 5M5,
Saskatchewan
Canada

Stewart, John
Emeritus Professor
118 Epron Road
Salt Spring Island, V8K 1C7, British
Columbia
Canada

Stone, John
3411 Paul Anka Drive, Unit 32
Ottawa, KIV 9R8, Ontario
Canada

Tiessen, Holm
IAI Director, c/o INPE
Av. dos Astronautas 1758
12227-010 São José dos Campos, SP
Brazil

Toledo, Peter Mann de
Rede Geoma/MCT
INPE/OBT
Av. dos Astronautas 1758
12227-010 São José dos Campos, SP
Brazil

Tourrand, Jean-François
University of Brasilia/CIRAD, SAS
Qd. 5, Bl. H, Edif IBAMA, 2° andar
70070-914 Brasilia, DF
Brazil

van den Belt, Marjan
Mediated Modeling Partners LLC
177 Ten Stones Circle
Charlotte, VT 05445
USA

Vernet, Maria
Integrative Oceanographic Division
Scripps Institution of Oceanography
University of California San Diego
La Jolla, CA 92093-0218
USA

Victoria, Reynaldo
CENA-USP
Av. Centenario 303
13416-970 Piracicaba, SP
Brazil

Vieira, Ima C. G.
Museu Paraense Emilio Goeldi
  (MPEG)
Av. Magalhães Barata 376
São Braz
66040-170 Belém, PA
Brazil

Weingart, Peter
Institute for Science and Technology
  Studies
University of Bielefeld
P.O. Box 10 01 31
D-33501 Bielefeld
Germany

Wood, Charles
University of Florida
Center for Latin American Studies
P.O. Box 115530
Gainesville, FL 32611-5530
USA

Woodrow, Maureen
Executive Director
Network Secretariat
Ocean Management Research
  Network
School of Management, Vanier 255f
University of Ottawa
Ottawa K1N 6N5
Canada

SCOPE Series List

SCOPE 1–59 are now out of print. Selected titles from this series can be down-loaded free of charge from the SCOPE Web site (http://www.icsu-scope.org).

SCOPE 1: *Global Environment Monitoring,* 1971, 68 pp
SCOPE 2: *Man-made Lakes as Modified Ecosystems,* 1972, 76 pp
SCOPE 3: *Global Environmental Monitoring Systems (GEMS): Action Plan for Phase I,* 1973, 132 pp
SCOPE 4: *Environmental Sciences in Developing Countries,* 1974, 72 pp
SCOPE 5: *Environmental Impact Assessment: Principles and Procedures,* Second Edition, 1979, 208 pp
SCOPE 6: *Environmental Pollutants: Selected Analytical Methods,* 1975, 277 pp
SCOPE 7: *Nitrogen, Phosphorus and Sulphur: Global Cycles,* 1975, 129 pp
SCOPE 8: *Risk Assessment of Environmental Hazard,* 1978, 132 pp
SCOPE 9: *Simulation Modelling of Environmental Problems,* 1978, 128 pp
SCOPE 10: *Environmental Issues,* 1977, 242 pp
SCOPE 11: *Shelter Provision in Developing Countries,* 1978, 112 pp
SCOPE 12: *Principles of Ecotoxicology,* 1978, 372 pp
SCOPE 13: *The Global Carbon Cycle,* 1979, 491 pp
SCOPE 14: *Saharan Dust: Mobilization, Transport, Deposition,* 1979, 320 pp
SCOPE 15: *Environmental Risk Assessment,* 1980, 176 pp
SCOPE 16: *Carbon Cycle Modelling,* 1981, 404 pp
SCOPE 17: *Some Perspectives of the Major Biogeochemical Cycles,* 1981, 175 pp
SCOPE 18: *The Role of Fire in Northern Circumpolar Ecosystems,* 1983, 344 pp
SCOPE 19: *The Global Biogeochemical Sulphur Cycle,* 1983, 495 pp
SCOPE 20: *Methods for Assessing the Effects of Chemicals on Reproductive Functions, SGOMSEC 1,* 1983, 568 pp
SCOPE 21: *The Major Biogeochemical Cycles and their Interactions,* 1983, 554 pp
SCOPE 22: *Effects of Pollutants at the Ecosystem Level,* 1984, 460 pp

SCOPE 23: *The Role of Terrestrial Vegetation in the Global Carbon Cycle: Measurement by Remote Sensing*, 1984, 272 pp

SCOPE 24: *Noise Pollution*, 1986, 466 pp

SCOPE 25: *Appraisal of Tests to Predict the Environmental Behaviour of Chemicals*, 1985, 400 pp

SCOPE 26: *Methods for Estimating Risks of Chemical Injury: Human and Non-Human Biota and Ecosystems, SGOMSEC 2*, 1985, 712 pp

SCOPE 27: *Climate Impact Assessment: Studies of the Interaction of Climate and Society*, 1985, 650 pp

SCOPE 28: *Environmental Consequences of Nuclear War*

Volume I: Physical and Atmospheric Effects, 1986, 400 pp

Volume II: Ecological and Agricultural Effects, 1985, 563 pp

SCOPE 29: *The Greenhouse Effect, Climatic Change and Ecosystems*, 1986, 574 pp

SCOPE 30: *Methods for Assessing the Effects of Mixtures of Chemicals, SGOM-SEC 3*, 1987, 928 pp

SCOPE 31: *Lead, Mercury, Cadmium and Arsenic in the Environment*, 1987, 384 pp

SCOPE 32: *Land Transformation in Agriculture*, 1987, 552 pp

SCOPE 33: *Nitrogen Cycling in Coastal Marine Environments*, 1988, 478 pp

SCOPE 34: *Practitioner's Handbook on the Modelling of Dynamic Change in Ecosystems*, 1988, 196 pp

SCOPE 35: *Scales and Global Change: Spatial and Temporal Variability in Biospheric and Geospheric Processes*, 1988, 376 pp

SCOPE 36: *Acidification in Tropical Countries*, 1988, 424 pp

SCOPE 37: *Biological Invasions: a Global Perspective*, 1989, 528 pp

SCOPE 38: *Ecotoxicology and Climate with Special References to Hot and Cold Climates*, 1989, 432 pp

SCOPE 39: *Evolution of the Global Biogeochemical Sulphur Cycle*, 1989, 224 pp

SCOPE 40: *Methods for Assessing and Reducing Injury from Chemical Accidents, SGOMSEC 6*, 1989, 320 pp

SCOPE 41: *Short-Term Toxicity Tests for Non-genotoxic Effects, SGOMSEC 4*, 1990, 353 pp

SCOPE 42: *Biogeochemistry of Major World Rivers*, 1991, 356 pp

SCOPE 43: *Stable Isotopes: Natural and Anthropogenic Sulphur in the Environment*, 1991, 472 pp

SCOPE 44: *Introduction of Genetically Modified Organisms into the Environment*, 1990, 224 pp

SCOPE 45: *Ecosystem Experiments*, 1991, 296 pp

SCOPE 46: *Methods for Assessing Exposure of Human and Non-human Biota SGOMSEC 5*, 1991, 448 pp

SCOPE 47: *Long-Term Ecological Research. An International Perspective,* 1991, 312 pp

SCOPE 48: *Sulphur Cycling on the Continents: Wetlands, Terrestrial Ecosystems and Associated Water Bodies,* 1992, 345 pp

SCOPE 49: *Methods to Assess Adverse Effects of Pesticides on Non-target Organisms, SGOMSEC 7,* 1992, 264 pp

SCOPE 50: *Radioecology after Chernobyl,* 1993, 367 pp

SCOPE 51: *Biogeochemistry of Small Catchments: a Tool for Environmental Research,* 1993, 432 pp

SCOPE 52: *Methods to Assess DNA Damage and Repair: Interspecies Comparisons, SGOMSEC 8,* 1994, 257 pp

SCOPE 53: *Methods to Assess the Effects of Chemicals on Ecosystems, SGOMSEC 10,* 1995, 440 pp

SCOPE 54: *Phosphorus in the Global Environment: Transfers, Cycles and Management,* 1995, 480 pp

SCOPE 55: *Functional Roles of Biodiversity: a Global Perspective,* 1996, 496 pp

SCOPE 56: *Global Change, Effects on Coniferous Forests and Grasslands,* 1996, 480 pp

SCOPE 57: *Particle Flux in the Ocean,* 1996, 396 pp

SCOPE 58: *Sustainability Indicators: a Report on the Project on Indicators of Sustainable Development,* 1997, 440 pp

SCOPE 59: *Nuclear Test Explosions: Environmental and Human Impacts,* 1999, 304 pp

SCOPE 60: *Resilience and the Behavior of Large-Scale Systems,* 2002, 287 pp

SCOPE 61: *Interactions of the Major Biogeochemical Cycles: Global Change and Human Impacts,* 2003, 384 pp

SCOPE 62: *The Global Carbon Cycle: Integrating Humans, Climate, and the Natural World,* 2004, 526 pp

SCOPE 63: *Alien Invasive Species: A New Synthesis,* 2004, 352 pp.

SCOPE 64: *Sustaining Biodiversity and Ecosystem Services in Soils and Sediments,* 2003, 308 pp

SCOPE 65: *Agriculture and the Nitrogen Cycle,* 2004, 320 pp

SCOPE 66: *The Silicon Cycle: Human Perturbations and Impacts on Aquatic Systems,* 2006, 296 pp

SCOPE 67: *Sustainability Indicators: A Scientific Assessment,* 2007, 448 pp

SCOPE Executive Committee 2005–2008

Index

Page numbers followed by *f* or *t* refer to figures or tables respectively.